TEXT BOOK OF PHYSICAL GEO[LOGY]

TEXT BOOK OF PHYSICAL GEOLOGY

GIRIJA BHUSAN MAHAPATRA
M.Sc, Former Geologist (GSI)

CBS Publishers & Distributors Pvt. Ltd.
New Delhi • Bengaluru • Chennai • Kochi • Kolkata • Mumbai
Hyderabad • Nagpur • Patna • Pune • Vijayawada

ISBN: 81-239-0110-0

First Edition: 1994
Reprint: 1999, 2000, 2001, 2002, 2004, 2008, 2009, 2010, 2011, 2012, 2013, 2014, 2015, 2016, 2017

Published by **Satish Kumar Jain** and produced by **Varun Jain** for
CBS Publishers & Distributors Pvt. Ltd.,
4819/XI Prahlad Street, 24 Ansari Road, Daryaganj, New Delhi - 110002
delhi@cbspd.com, cbspubs@airtelmail.in • www.cbspd.com
Ph.: 23289259, 23266861, 23266867 • Fax: 011-23243014

Corporate Office: 204 FIE, Industrial Area, Patparganj, Delhi - 110 092
Ph: 49344934 • Fax: 011-49344935
E-mail: publishing@cbspd.com • publicity@cbspd.com

Branches:
- ***Bengaluru:*** 2975, 17th Cross, K.R. Road, Bansankari 2nd Stage, Bengaluru - 70 • Ph: +91-80-26771678/79 • Fax: +91-80-26771680 E-mail: cbsbng@gmail.com, bangalore@cbspd.com
- ***Chennai:*** No. 7, Subbaraya Street, Shenoy Nagar, Chennai - 600030 Ph: +91-44-26681266, 26680620 • Fax: +91-44-42032115 E-mail: chennai@cbspd.com
- ***Kochi:*** Ashana House, 39/1904, A.M. Thomas Road, Valanjambalam, Ernakulum, Kochi • Ph: +91-484-4059061-65 Fax: +91-484-4059065 • E-mail: cochin@cbspd.com
- ***Kolkata:*** 6-B, Ground Floor, Rameshwar Shaw Road, Kolkata - 700014 Ph: +91-33-22891126/7/8 • E-mail: kolkata@cbspd.com
- ***Mumbai:*** 83-C, Dr. E. Moses Road, Worli, Mumbai - 400018 Ph: +91-9833017933, 022-24902340/41 • E-mail: mumbai@cbspd.com

Representatives:
- Hyderabad: 0-9885175004
- Nagpur: 0-9021734563
- Patna: 0-9334159340
- Pune: 0-9623451994
- Vijayawada: 0-9000660880

Printed at:
Neekunj Print Process, Delhi (India)

To my loving elder borthers

`SANTOSH BHAINA

&

BIBHUTI BHAINA'

PREFACE

My basic objective in writting this book "Text Book of Physical Geology" has been to enable the reader to apprehend fully the fundamental principles and processes of Physical Geology and also to meet the requirements of the students who have the interest and intention to acquire more knowledge in the field of physical geology. I shall consider my efforts successful if the present book satisfies the need of the students of our Universities. Suggestions for the improvement of the book are welcome.

The whole credit of my writing this book goes to my mother and both of my elder brothers for their timely encouragements. I am highly indebted specially to Bibhuti - bhaina for the sincere guidance and help as well as for the pain he had taken without which perhaps it wouldn't have been possible for me to complete the manuscript.

My sincere thanks go to Shri Sarat Chandra Samal one of my close associates who had rendered valuable suggestions and co-operation in my efforts for writing the book.

I am thankful to the M/s CBS Publishers & Distributors Pvt. Ltd. and M/s Super Computer's for the timely publication of the book in a presentable form.

G.B. Mahapatra

CONTENTS

1

GEOLOGY: SCOPE AND IMPORTANCE

The word Geology has been derived from the Greek words 'Ge' meaning the earth and 'logos' meaning science. Thus, geology is the scientific study of the earth. The study of geology mainly concerns itself with the study of the earth's constitution, structure and history of development as well as the outer solid shell of the earth composed of rocks, which is known as lithosphere.

Geology is a science of many facets and includes the study of :

1. *Physical Geology* It deals with the endogenous (internal) and exogenous (external) agencies and the processes that bring about changes on the earth's surface. James Hutton is regarded as the father of physical geology.

2. *Geo Tectonics* It concerns with the movements of the earth's crust and the deformations caused by them.

3. *Structural Geology* It deals with the configuration of the rocks in the earth's crust produced due to a number of forces generated both exogenously and endogenously.

4. *Geomorphology* It deals with the study of landforms.

5. *Crystallography* It is the study of the external forms and internal atomic structure of the crystalline minerals.

6. *Mineralogy* It deals with the minerals, their composition, characteristics, modes of occurrence and origin.

7. *Petrology* It deals with the origin, structure, texture, mineralogical composition etc. of the different types of rocks.

8. *Stratigraphy* It deals with the strata of sedimentary rocks, their succession, thickness, age, variations and correlations. Thus it is the study of strata as a record of geological history.

9. *Palaeontology* (Greek-'Palaios' meaning ancient and 'Ontos' meaning being). It is the study of fossils of plants and animals that are found in the rocks of past geological periods. They indicate the climate, age and environment of deposition of the rock unit in which they are found.

10. *Economic Geology* It deals with the study of mineral deposits, their modes of formation, modes of occurrence, distribution etc.

11. *Engineering Geology* It deals with the application of geological knowledge in the field of engineering for the construction of dams, bridges, tunnels, buildings, roads along hill slopes etc.

12. *Hydrology* It deals with the hydrological properties of rocks and the occurrences of ground water, its movement and action.

13. *Geophysics* It is a branch of geology with the application of physics which includes geodesy, seismology, meteorology, oceanography and terrestrial magnetism.

14. *Geochemistry* It deals with the chemical constitution of earth, the distribution and migration of various elements in various parts of the earth.

15. *Mining Geology* It deals with the application of geology in the mining and extraction of minerals.

The knowledge of geology is of exceptionally great practical value. The mineral resources of the earth have been used to a certain extent since pre-historic time. Modern civilisation is largely dependent on minerals and mineral products. Mining has already become one of the leading industries in the world. The ores are exhausted with continous mining and in order to extract the ores man is compelled to prospect new areas to find unexploited minerals. Geology forms the theoretical basis for prospecting and for the exploration and working of all mineral deposits without exception.

The location of suitable sites for dams, buildings, tunnel construction, roads as well as the protection of coastal areas from erosion, flood control measures, exploration of ground water etc. are only successful with the application of geological knowledge.

Geological knowledge concerning soils, erosion, drainage and mineral fertilisers has wide application in the field of agriculture also.

Thus the study of geology has too much utilitarian significance in the present day civilisation

2

THE EARTH—IT'S POSITION IN SPACE

The Earth, as we know, is a member of the planetary system of the Sun. Besides the Sun, the solar system includes nine planets, their satellites, asteroids and comets. The planets are located in the following order away from the Sun: Mercury, Venus, Earth, Mars, Jupiter, Saturn, Uranus, Neptune and Pluto. The Earth is situated approximately 150 million kilometres away from the Sun.

Before Copernicus demonstrated that the Sun is the centre of our solar system, the 'Geocentric Theory' regarding the Earth was getting wide acceptance. According to this theory, the Earth was the unmoving centre of the universe, round which the Sun and the stars and all other heavenly bodies revolved. The 'Heliocentric Theory' evolved by Copernicus stated that the Sun was the centre of the universe and that the Earth and other planets revolved round it. The belief that Sun was the centre of the universe was corrected by the German astronomer Johan Kepler in 1609. Galileo Galilei, the Italian astronomer and Sir Isaac Newton, the German scientist, had conclusively established that the Sun is the centre of the solar system. Modern theories on the Earth are based on the Copernican theory.

All the planets of the solar system are divided into two groups. The inner planets i.e. those of the terrestrial group, which are nearer to the Sun and are the denser members of the solar system and include Mercury, Venus, the Earth and Mars. The outer planets are bigger in volume and lighter in density. They include Jupiter, Saturn, Uranus, Neptune and Pluto.

The Earth, the planet on which we live has the highest density among terrestrial group of planets. It has one satellite, the Moon. The Earth is unique amongst the planets of the solar-system in possessing a complex variety of living forms and an atmosphere of a nitrogen composition.

The following are some of the important facts about the Earth:

1. The shape of the Earth is that of a spheroid.
2. Its equatorial radius is 6378.3 km.
3. Its polar radius is 6356.9 km.
4. Mean radius is 6371.2 km.
5. Mass of the Earth = 5.975×10^{27} gm.
6. Volume of the Earth = 1.08×10^{27} cc.
7. Average density = 5.5 gm/cc.
8. Average density of surface rocks is nearly 2.8 gm/cc.
9. Age of the Earth = 4.5 billion years.
10. Area of the Earth's surface = 510.08 million square km.
11. Total Land-surface = 148.63 million square km.
12. Total water-surface = 361.45 million square km.
13. The Earth completes one full rotation around its axis in 23 hours 56 minutes.
14. The Earth completes a full revolution around the Sun in 365.26 days.

ORIGIN OF THE EARTH

Since the Earth is a member of the solar system, it is commonly believed that the origin of the Earth is connected with that of the solar system. All the principal theories which have been advanced to explain the origin of the Earth, have in common the idea that the planets evolved from the Sun. Regarding the origin of the Earth a number of theories have been put forward but none of them can be said to be perfectly correct. Some of the important theories explaining the origin of the earth are as follows:

1. The earliest theory about the origin of the earth was advanced by the French philosopher Georges Buffon in 1745. According to this theory material was pulled out of the Sun by an external force such as gravitational pull resulting from the near collision of the Sun with another star. The cooling of the blobs of solar matter ejected from the Sun during the cataclysmic collsion gave rise to the planets of which the Earth is a member.

2. *Nebular Hypothesis* The first Nebular Hypothesis was advanced by the German philosopher Kant (1755) and then by the French mathematician Laplace (1796). According to this hypothesis, the planetary system is believed to have been evolved from a large, hot, gaseous nebula rotating in space (Latin 'nebula' means mist). The rotating nebula, according to the law of universal gravitation, became more compressed and compact with an increase in the speed of rotation. Gradual cooling with contraction in size and increasing concentration of mass towards the centre of the nebula led to an increase in the rate of rotation and a growth of centrifugal force. With the increased velocity of rotation, the centrifugal force around the equator of the mass eventually became equal to the gravitational attraction between the material at the outer rim of the disc and the central mass. As a result a ring of material was left while contraction of the remaining material continued. When the cen trifugal force exceeded the force of gravity in the equatorial zone of the nebula a ring of matter began to spin off along the whole periphery of the rotating disc. Thus successive rings of matter were formed and left behind the contracting mass. Further

cooling and coalescence of the rings led to the formation of planets and their satellites, while the remnant of the pre-existing nebula formed the central incandescent mass of the solar system and is known as the Sun. According to the Kant—Laplace hypothesis, the earth was originally incandescent and in the process of its development it became cooled and contracted. (see fig)

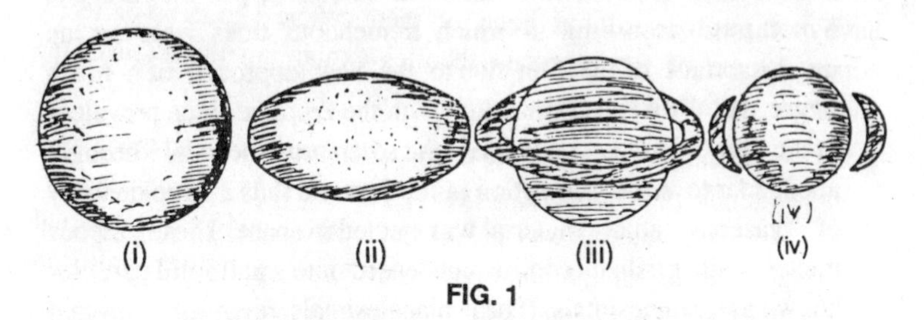

FIG. 1

Merits of the Nebular Hypothesis (i) This hypothesis is able to explain for similar directions of rotation of all the planets and the Sun.

(ii) It also accounts for the same plane of rotation of the planet.

Demerits of the Hypothesis (i) It does not satisfy the principle of conservation of angular momentum in the solar system. This theory was rejected when it was learnt that the angular momentum of the solar system is concentrated in the planets and not in the Sun. This is not compatible with the idea that the mass of the matter rotated more rapidly as it condensed.

(ii) Certain planets, for e.g. Venus and some satellites of Mars and Jupiter etc. rotate in a direction opposite to that of the other planets and the Sun.

(iii) If this hypothesis is assumed to be correct then probably there would have been the formation of another ring or a planet.

(iv) The process of condensation of highly rarified gas into rings rather than its dissipation into space is questioned.

(v) The hypothesis fails to explain the observed differences in the density, size and mass of the planets of the solar system.

3. *Planetesimal Hypothesis* This is a type of "Dynamic Encounter Theory" which believes that material is pulled out of the Sun by an external force such as gravitational pull resulting from the dynamic encounter of the Sun with another star. Chamberlin and Moulton (1904), two American scientists, put forward this hypothesis; according to which tremendous tides were set up on the surface of the Sun due to the near approach of a much larger star. This, in conjunction with the eruptive force prevalent in the Sun (which is known as the solar-prominences) brought about a large scale disruption of the Sun and thus a large quantity of gaseous solar material was ejected in space. These gaseous matter with gradual cooling condensed into small solid particles known as planetesimals. These planetesimals revolved around the Sun in different orbits, during which many of them collided with one another, thereby the small planetesimals aggregated to form planets.

Merits (i) It explains the occurrence and occasional fall of mete orites.

(ii) It also explains the prevalence of highly elliptical orbits with the smaller planetary bodies and nearly circular orbits of the larger ones.

Demerits (i) It fails to indicate how gaseous material drawn out from the Sun condensed to form planetesimals instead of being dispersed through out the space.

(ii) The probability of near approach of another large star to the Sun seems quite remote and there is no evidence to support this assumption.

Thus this theory never found much support.

4. *Tidal Hypothesis* Jeans (1919), a British astronomer and Jeffreys (1929) had propounded a modified theory of biparental origin of the solar system. This hypothesis states that the planets were formed from a gaseous filament that was torn out from the Sun due to the gravitational pull of a huge star which

happened to approach very close to the Sun. This large filament of gas thus ejected in space was extremely unstable which immediately got splitted up into a number of fragments. These fragments with gradual cooling and condensation gave rise to the planets and thus the solar system came into existence.

Merits (i) The rotation of the gaseous filament accounts for the concentration of angular momentum in the planets rather than in the Sun itself.

(ii) This is a simple hypothesis.

Demerits (i) The process of formation of planets seems quite mysterious.

(ii) Encounter of one star with the Sun does not appear to be correct. Since the stars and the Sun are so apart from each other that passing of one star near the Sun does not appear to be convincing.

(iii) The ejected gaseous matter would have been dispersed into the space, instead of forming planets by condensation.

5. *Double Star Hypothesis* This hypothesis was put forward by Lyttleton in 1938. According to this hypothesis, it is believed that prior to the evolution of the solar system, alongwith the Sun there existed another companion star which was captured later on by another huge star that came very close to the Sun and its companion star. While the huge star started receding away the filament drawn by the attraction of the invading star came into the control of the Sun which later gave rise to the planets etc. in a similar way as explained by Jeans and Jeffreys in their 'Tidal Hypothesis'.

6. *Weizsacker's Hypothesis* In 1944, a German physicist, C.F. Von Weizsacker proposed a modification of the Nebular Hypothesis. According to this hypothesis, the Sun was surrounded by a thin, flat, rapidly rotating cloud of matter that encircled its equator. This cloud was having a much larger quantity of hydrogen and helium which had escaped into space in course of time because of their being extremely light in weight. The materials that formed the planets were carried as floating dust

particles in the rotating gaseous envelope round the Sun and by gradual accretion of these matter the planets were formed. This hypothesis suggests a cold-origin of the earth.

Merits (i) In the words of E.W. Spencer-"Radiation from the Sun drove off most of the hydrogen and helium which are extremely light in weight. The heavier elements left behind collided with the escaping hydrogen and helium and angular momentum was transferred to them. This accounts for the concentration of momentum in the planets".

(ii) This theory explains the distribution and differences in the density and size of the inner and outer group of planets.

This hypothesis is also not considered to be the most satisfactory scientific explanation for the origin of earth.

7. *Meteorite Hypothesis* This hypothesis was suggested by the Soviet scientist Otto Schmidt in 1944. According to this hypothesis, the Earth and the other planets of the solar system were formed from the cluster of interstellar matter captured by the Sun during its passage near the centre of gravity of the galaxy. These are protoplanetary bodies with their own angular momentum. Through accretion of these matter, they began to grow in size and mass and exerted gravitational pull on one another and ultimately gave rise to planets and satellites.

   Schmidt has mathematically substantiated the existing regularity in the distances between the planets of the solar system and their actual distribution by mass and density. Besides, this hypothesis is also able to explain the shape of the planetary orbits, their direction of rotation etc.

   The drawback of this hypothesis is that the Sun's capturing the cluster of interstellar matter does not appear convincing.

8. *Proto-planetary Hypothesis of Kuiper* G.P. Kuiper, an American scientist (1951) proposed a modification of Weizsacker's Hypothesis. He suggested that there existed a flattened and slowly rotating disc-shaped solar nebula bulging out from the equator of the Sun. This disc-shaped nebula became internally unstable and broke up into smaller concentrations called "proto-planets", which later on collided and united to form bigger masses which were the planets.

9. *Fred Hoyle's Magnetic Theory* In 1958 Fred Hoyle propounded his hypothesis, according to which the protoplanetary cloud was created in the process of differentiation of the Sun from an original nebular matter that had been undergoing contraction. The differentiation of the nebular matter into the Sun and the gaseous cloud was due to fast rotation of the nebular mass. The process of differentiation, however, came to an end due to magnetic coupling between the Sun and the gaseous cloud. The aggregates of the particles in the gaseous cloud, gave rise to planets.

   But this hypothesis did not find much acceptance.

10. *V.G. Fresenkov* (1960) propounded that the Sun and its planets were formed through the condensation of matter with in a galactic cloud. This cloud consisted of hydrogen, helium and a small quantity of heavier material. At first, the Sun took shape at the core of the cloud (globule) and its evolution was accompanied with repeated ejection of material into the protoplanetary cloud with the result that the Sun lost some of its mass and transferred a considerable fraction of its angular momentum to the forming planets. Thus, the Sun and the planets were formed from different parts of one and the same original matter.

    Now-a-days the 'Big-Bang Theory" is getting wide accep tance, according to which large scale explosion with in the nebula has given rise to the planets and the Sun. The correctness of this theory is yet to be proved.

Conclusion

All ideas concerning the origin of the Earth and the solar system have their problems, and new discoveries often add to the demerits to the theories.Thus,origin of the solar system continues to be a problem and even the most modern theories contain many points that need verification.

3

AGE OF THE EARTH

The age of the Earth was a matter of speculation till very recent times and as such there was divergence of opinions about the antiquity of the Earth. Until recently geology relied extensively on the concept of the relative age of rocks. It was only with the discovery of radioactivity that a new method giving an approximate age, with comparatively less chances of errors, was found.

The determination of the age of the earth was attempted through two distinct processes:

1. Indirect methods for ascertaining the earth's age, and
2. Radioactive methods for determining the actual age (Direct-method).

1. *Indirect Methods*

(a) *Sedimentation-Clock* This takes into account the average annual rate of sedimentation and the thickness of all strata deposited during the whole geological history. Although this method is full of imperfections and variations from place to place, the age of the Cambrian as determined by this and other sophisticated methods closely approach each other.

According to this method, there is an average rate of deposition of 30 cm of sediments in 755 years. As such the beginning of Cambrian sedimentation come to about 510 million years.

(b) *Salinity Clock* Joly and Clarke took into account the rate of accumulation of sodium in ocean-water for determining the age of the Earth. The yearly rate of increase in the sodium content (salinity) of sea water per unit volume can be determined from direct observation and the content of sodium per unit volume of ocean water can also be calculated. Accordingly, the total amount of sodium in the ocean water can also be determined from the total volume of ocean water.

This method has established that the Earth was formed about 100 million years ago. But this method also suffered from a lot of imperfections, for e.g . the rate of supply of sodium to ocean is not uniform throughout the geological history.

(c) *Evolutionary Changes of Animals*, The evolution of life has proceeded since the first forms of life appeared on earth. As we know, the first formed animals were. unicellular which under went various phases of the evolutionary processes and multicellular organisms with more complexities came into being. Man is, accordingly, considered to be the most evolved one.

Evidence of the process of evolution, is found in the form of fossils, (preserved parts of animals and impressions of plants and animals). Taking into account the evolutionary developments biologists have estimated the age of the earth to be about 1,000 million years.

(d) *Rate of Cooling of the Earth* Kelvin estimated the age of the earth on the basis of his study of the history of cooling of the earth. Since temperature increases with depth, Kelvin assumed that the earth began as a molten mass and has been cooling ever since i.e. the earth is progressively losing heat. Assuming the initial temperature of the earth to be 3900°C, Lord Kelvin estimated the age of the earth to be 100 million years and later on revised it to be between 20 to 400 million years.

This method of estimation is full of imperfections and one of the most important consideration regarding the generation of radio-active heat was not taken into account in this method.

(e) *Varved Sediments* The term varve is applied to glacial-lake deposits containting rhythmically laminated sediments of clay and silt. The silt and clay laminae occur alternate with each other in a regular pattern. The couplet of beds is recognised as representing a year, like the annual rings of a tree, and is called a varve. It is presumed that these so called varved clays accumulated in lakes as a result of thawing of glacier-ice; the fine bands having been deposited in winter and the coarser ones, in summer. A varve is usually about 10 mm in thickness and occasionally it ranges up to 300 mm. By counting the number of pairs of bands in a given section, the time represented by the section can be ascertained in years. Geologic time ranging from '0' to 10,000 years could be counted on this varve clock.

Apart from the above indirect methods, other attempts for estimating the age of the earth were made on the basis of contraction of the earth's surface due to cooling, by the rate of limestone deposition etc. Even, on the basis of the concept of separation of the moon from the body of the earth, Charles Darwin estimated the age of the earth to be 57 million years.

2. *Radioactive Method* The discovery of radioactivity brought a new and apparently precise concept of measuring the age of the earth. Radioactivity, as we know, consists in the spontaneous disintegration of the nucleus of unstable elements and their transformation into stable isotopes or new elements. The disintegration is accompanied with the emission of alpha particles (i.e. helium nuclei), beta particles (electrons) and energy in the form of gamma radiation. The basic principle underlying all the radioactive methods is that "a radio-active parent element decays into a stable daughter element at a constant rate". The radioactive decay is usually expressed in terms of "half life period'. A half-life period is the time required for one half of an original amount of a radioactive element to disintegrate. For geological purposes the unit

of time is one year. The relationship between the half life period and the rate of decay has been found out to be as:

$T = 0.693/\lambda$, where T = half- life period λ = rate of decay

Usually the 'half life period' is determined and accordingly it is equated to find out the age of the earth.

The following are some of the common methods used for the purpose of determining the age of earth:

(a) *Uranium-lead method* Here two isotopes of uranium are used, U^{238} and U^{235}. The chemical element uranium, a heavy metal of atomic weight 238 spontaneously gives away X-rays and atoms of the helium gas and is ultimately converted into the chemical element-radium of weight 236. This element too continues emitting helium atoms till it is reduced to lead of atomic weight 206.

$$U^{238} \longrightarrow Pb^{206} + 8He^{4}$$

The half life of Uranium 238 is 4500 million years. One gram of uranium-238 will produce 1/7600,000,000 gram of stable lead.

Similarly the second isotope U^{235} undergoes sponta neous disintegration and ultimately gives rise to lead of atomic weight^{207}.

$$U^{235} \longrightarrow Pb^{207} + 7He^{4}$$

The half life of U-235 is 713 million years.

(b) *Thorium lead method* Thorium-232 through radioactive disintegration gives rise to lead 208. The half life period, in this case is 13,900 million years.

$$Th^{232} \longrightarrow Pb^{208} + 6He^{4}$$

(c) *Potassium argon method* Potassium, an element present in many minerals and rocks, have three isotopes K^{39}, K^{40} and K^{41}. Only K^{40} is radioactive. The radioactive transformation of $K\text{-}^{40}$ consists in absorbing the electron by the nucleus with the nearest to its electronic shell. Thus, a radiogenic stable isotope of argon with exactly the same atomic weight (Ar^{40}) is formed. Only 12.4% of K^{40} through this process is converted to Ar^{-40}. The half life period, here is 11900 million years.

$$K^{40} \xrightarrow{\text{electron capture}} Ar^{40}$$

About 87.6% of K^{40} through emission of beta particles get converted into Ca^{40} (calcium-^{40}) with a half life period of 1470 million years. But this method does not find much application in the determination of the age of earth. It is mainly due to the fact that most common rock forming minerals already contain so much of primary calcium (Ca^{40}) that the compara tively minute amounts of radiogenic Ca^{-40} cannot be determined.

(d) *Rubidium Strontium method* This method is based on the radioactive decay of Rb^{87} and its transformation in to Sr^{87}. In minerals Rb-atoms are associated with its decay product as well as the common Sr^{86}. Natural rubidium is having two isotopes Rb^{85} and Rb^{87}. Rb^{87} is radioactive. The ratio of Sr^{87} to Sr^{86} can be determined with great accuracy so also the ratio of Rb^{87} to Sr^{87}. It is therefore, this method is more reliable for metamorphic rocks. The half life period is 50,000 million years.

$$Rb^{87} \xrightarrow{\text{Beta particles}} Sr^{87}$$

(e) *Radio Carbon Method* Cosmic rays, at the upper atmosphere, change nitrogen (N^{-14}) to Carbon (C^{-14}), an isotope of carbon 'C^{12}'. Once formed carbon-^{14} is quickly disseminated through the atmosphere and reacts with oxygen to form CO_2 which is absorbed by all living matter. It has been observed that a constant level of this isotope (i.e C^{-14}) is maintained by all living organisms. At death, the organism stops absorbing the carbon-14 and the carbon-14 present in the organism starts decreasing at a constant rate. The half life period of C-14 is about 5,730 years. Since the half life of radioactive carbon and its present content are known, it is possible to establish the time when the organism died and, therefore, the age of the rock. This method is especially useful for dating relatively recent materials up to 70,000 years.

CONCLUSION

The best means of estimating the age is provided by the Uranium- lead, Rubidium-strontium and Potassium-argon methods. On the basis of the data provided by the radioactive methods described above, the age of the earth has been estimated to be of 4500 million years i,e 4.5 billion years.

4

INTERNAL STRUCTURE OF THE EARTH

Direct observation of the interior of the earth is not possible due to the fact that the interior becomes hotter with depth, which is convincingly indicated by the volcanic eruptions. Besides, the deepest hole in the earth, a drill hole, is only about 8 km deep; this is quite negligible in comparison with the radius of the earth. Since direct observation of the interior of the earth can not be made conveniently, all the important sources of data on the structure of the earth are indirect and are logically derived and inferred from other evidences. Apart from the seismological studies, other important sources of data like meteorite analysis etc., even though indirect, logically prove that the earth's body comprises of several layers which are like shells resting one above the other. These layers are distinguished by their physical and chemical properties, particularly in their thickness, depth, density, temperature, metallic content etc.

Now-a-days much information about the interior of the earth has been obtained from the study of the propagation of the earth-quake waves through the earth. As it has has been identified, there are three types of earthquake waves:

(i) Longitudinal, Primary or P-waves. They are similar to sound waves in which the particles move to and fro in the direction in which the wave is travelling. These waves travel in solid, liquid and gaseous media. They have short wave length and high frequency.

(ii) Transverse, Secondary, S-waves or Sheer waves. These waves are like the waves which run along a string which is fastened at one end, stretched fairly tight and shaken at the other end. In such waves the particles move to and fro at right angles to the path of the wave. These waves travel only in solid medium. In comparison to the Primary waves, they are slow in motion. They also have short wavelength and high frequency.

(iii) Surface, L -waves, Rayleigh or R-waves. These are transverse waves and are confined to the outer skin of the crust and are responsible for most of the destructive force of earthquakes. They have low frequency, long wave length and low velocity.

The P and S-wave velocities change with depth and each change can be related to a change in materials. Each region of changing shock-wave velocity demarcates a zone of discontinuity. This also gives the following information about the interior of the earth:

(i) the shells of increasing density are found towards the centre of the earth and it is estimated to be 18 at the centre.

(ii) each shell is formed of different materials i.e the materials vary in their chemical composition, physical properties as well as in their state (i.e. solid, liquid or gas etc.). These changes attribute to the existence of minor discontinuities.

As à whole, on the basis of seismic investigations, the earth's interior has been broadly divided into three major parts:

1. The Crust

2. The Mantle

3. The Core

It has also been inferred that —

(a) the crust, mantle and core are separated by two sharp breaks, usually known as major discontinuities;

(b) the crust is having an average thickness of about 33 kms;

(c) the crust is composed of heterogeneous materials;

(d) the second major segment of the earth, i.e the mantle extends from below the crust to a depth of 2900 kms;

(e) the third major segment of the earth i.e. the core extends from below the mantle upto the centre of the earth;

The Crust It is the uppermost shell of the earth that covers the rocks of the interior thinly. Its thickness over the oceanic areas is generally 5 to 10 kms; whereas on the continental areas it is about 35 kms and the thickness ranges from 55 to 70 kms in orogenic belts. The Mohorovicic discontinuity marks its lower boundary.

From the study of the shallow focus earthquakes and artificial seismic explosions, it has been inferred that there are two zones of crustal rocks beneath the continents, although only one occurs beneath the oceans. In the continental regions, underneath a zone of superficial sediments, the crust can be divided into two layers, the upper layer called 'Sial' and the lower one 'Sima'. The boundary between the sial and the sima is called the Conrad Discontinuity.

(i) *Sial* It is also known as the upper continental crust. It consists of all types of rocks, igneous, sedimentary and metamorphic, which are exposed at the land surface. This layer is rich in silica and aluminium. The rocks in this layer are of granitic to grano - dioritic composition.

The Conrad discontinuity which is located at a depth of 11 kms separates the sial layer from the underlying sima layer.

(ii) *Sima* It is also known as Lower Continental Crust. Its thickness is about 22 kms. It extends from the Conrad discontinuity up to the Mohorovicic discontinuity. This layer is rich in silica and magnesium and is basaltic in composition. It includes two parts:

(a) Outer Sima and (b) Inner Sima

Outer sima extends upto a depth of 19 kms and consists of rocks of intermediate composition.

Inner sima is located at a depth of about 19 kms and extends upto 33 kms. It is basic to ultrabasic in composition.

It has been observed that the L-waves while passing through ocean floors acquire more velocity in comparison to its propagation through the land masses. This indicates that the sialic layer (granitic material) with which the landmasses are usually composed of, is practically absent on the ocean floors. The ocean floor is of basaltic composition, which are poorer in potassium and richer in aluminium than the basalts of the land surface and are called 'oceanic tholeiites'.

A change from continental to oceanic crust takes place at the peripheries of the major continents where there are marginal seas and island arcs.

The Mantle The second major part of the earth is the mantle which is the source-region of most of the earth's internal energy and of forces responsible for ocean-floor spreading, continental drift, orogeny and major earthquakes. The mantle extends from below the Mohrovicic discontinuity (which separates it from the overlying crust) up to a depth of 2900 kms. Thus, its thickness is about 2865 kms. It forms about 83 per cent of the earth by volume and 68 per-cent by mass.

Since the P and S waves record a definite increase in their velocities with depth, it is logically assumed that the material of the mantle is more dense than that of the overlying crustal rocks. The material is olivine-pyroxene complex , which exists in a solid state. It is believed that the upper mantle has a mix of 3 parts of ultramafic rocks and one part of basalt. This mix is known as Pyrolite.

The upper-mantle extends upto a depth of 1000 kms. The lower mantle extends from 1000 kms to the core boundary (Gutenberg-Weichert discontinuity).

The *upper mantle* is consisting of two layers which are distinguished on the basis of velocity of propagation of seismic waves. The upper layer of the upper mantle lying between Mohorovicic discontinuity and a boundary at a depth of 410 km is characterised by a decrease in the seismic velocity. This layer is called the Gutenberg layer. The crust and the upper part of the Gutenberg layer together constitute what is known as lithosphere. The lithosphere is underlain by 'asthenosphere' which is a layer of virtually of no strenghth to resist deformation and it is the low seismic velocity layer. The asthenosphere is situated somewhat between 70 to 220 km depth. To be more precise, the lithosphere is separated from the rest of the mantle by the asthenosphere.

The lower part of the upper mantle is known as Golitsyn's layer in which the velocity of the seismic waves sharply rise, reaching about 11.3 to 11.4 km/second at the depths of 900-1000 km.

The lower-mantle is about 1900 kms thick, and consists of two parts: (i) 1000 kms to 2700 kms and (ii) 2700 kms to 2900 kms. The upper layer is characterised by a further increases in the seismic-velocity. The velocity of P-wave reaches its maximum i.e 13.7 km/sec at this layer. At a depth of 2700 to 2900 kms, the velocity of propagation decreases to 12.6 km/second for the primary-waves. This may

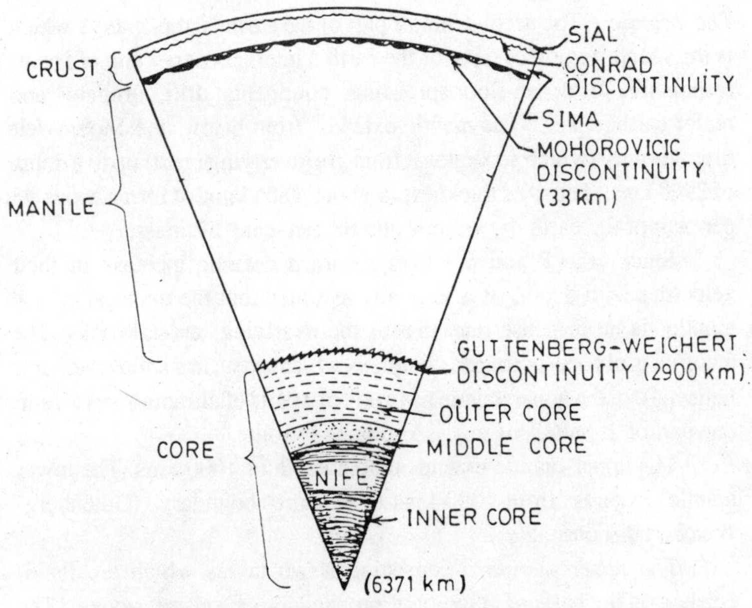

FIG.2. [DIVISION OF EARTH'S INTERIOR]

probably be due to the existence of a transitional layer.

Within the mantle a number of second order discontinuities have been located. (The first order discontinuities are the major ones separating the crust from the mantle and the mantle from the core i.e. the Mohorovicic Discontinuity and Guttenberg-Weichert Discontinuity).

(i) *Density break* At a depth of 80 km, density changes from 3.36 to 3.87.

(ii) *Gravity break* At a depth of 150 km, gravity changes from 984 cm/sec^2 to 974 cm/sec^2 till it reaches a depth of 1200 km.

(iii) *Seismic-discontinuity* At 410 km depth, there is a decrease in the seismic velocity.

(iv) At 700 km. depth, there changes the capability of the materials in storing the elastic-strain energy. Up to 700 kms. the capability is more.

(v) *Repetti discontinuity* At 950 to 1000 km depth there is a rapid rise in the velocity of seismic waves.

(vi) *Gravity break* At 1200 km depth, gravity attains its minimum value i.e. 974 cm/sec^2, thereafter it rises up to 1068cm/sec^2 at the core-boundary.

(vii) *Seismic-discontinuity* At a depth of 2700 km. the seismic velocity reaches its maximum i,e, 13.7 km/sec and gradually it decreases.

The Core It is the innermost part of the earth. It is separated from the mantle by the Guttenberg-Weichert discontinuity and extends upto the very centre of the earth. It constitutes around 17 percent of the volume and 34 percent of the mass of the earth.

Since the S-waves do not pass through the outer core, no information about the inner core is provided by them. It only suggests that the core, at least in its outer part, is fluid-like in its character because it does not transmit the S.waves and because it retards the velocity of P-waves from 12.6 to 8.4 km/sec.

The pressure and temperature in the core are both very high. The pressure is assumed to be over three million atmosphere and the temperature is around 6000°C . There is a sharp change in the density from about 5.5 x 10^3kg m^{-3} in the mantle to about 10.6 x 10^3 kg m^{-3} in the core, while at the centre of the core the density increases to 12 or 13x10^3 kg m^{-3}.

The core consists of three parts:

(i) *Outer-core* (ii) *Middle-core* and (iii) *Inner-core*.

(i) *Outer core* It extends from 2900 km to 4982 kms. It is considered to be in a state of homogeneous fluid. It does not transmit S-waves.

(ii) *Middle core* It is a transition layer, that extends from 4982 kms to 5121 kms. The material is in a fluid to semifluid state.

(iii) *Inner core* It extends from 5121 kms upto the centre of the earth i.e. 6371 kms. The inner core is assumed to be in a solid state, with a density of about 13. It is believed to contain metallic nickel and iron and is called 'Nife' Its thickness is about 1250 kms.

From the study of the geochemical differentiation of the earth, it has been indicated that in its early stages of development the planet must have been in a liquid state. The study of meteorites, the distribution of elements in the whole substance while an ore is treated in a blast furnace etc. give evidences to assume that the earth is having a nickel-iron core. It is presumed that when all the oxygen and sulphur had been consumed in reactions with the active metals, excess iron would separate in the form of droplets. Because of their comparatively high density these drops of iron would sink through the molten or viscous silicates towards the centre of the planet, eventually concentrating there into a central metallic mass, the core of the earth. Several of the rarer metals which are not chemically very active, gold, platinum, nickel, for example would react to only a moderate extent with oxygen and sulphur. The excess free rare metals would tend to concentrate in the descending iron droplets and would also ultimately find home in the core. This presumption goes a long way to explain the nickle-iron (nife) composition of the core of the earth. The exact nature of the core is still controversial.

A recent discovery about the earth's core as published in 'The Times of India (9-6-1988) reads as follows:

Lakes of Molten Rock in Earth's Core

Washington June 8. (PTI). Scientists have now found that the supposedly smooth hot ball of iron at the earth's core is not smooth but has vast mountain-range sized bumps and valleys with upside down lakes of molten rocks between them.

Other such startling discoveries have been made by scientists who can now see deep inside the earth. Using for over past two years seismic tomography, which produces X-ray like images down to the core of the planet, they can photograph for the first time objects deep under th ground.

This technique will help to solve mysteries, that have scientists for decades, such as why satellites do not fly sm orbit but bob up and down like corks on water, risir hundreds of feet on each tour, and why the day varies in hic spins jerkily on its axis.

Analysing these recent findings the "Wology, the an article that the answers to these small puzz of California such as what makes the continents move, new tomographic are just beginning to emerge as tomography.

Harvard, the Massachearth's core, a 4000-mile-California Institute of Tecthe perfect sphere as depicted at San Diego have tThe tomography images show vast technology. seven miles high and deep. which are

One recen are made of rock while the core inside them diameter ball

in geology mountaire upside down in relation to the earth's surface, the upsithe surface of the earth's core have been dubbed "anti-its" and "anti-oceans."

Some researchers believe that these mountains have oceans of lighter density iron between them, thus making the upside down array somewhat like anti-continents and anti-oceans 1800 miles deep.

At the centre of the core the pressure is so great that not even temperatures of 12,000 degrees fahrenheit can keep the iron liquid. It is compressed into a solid again, in hexagonal crystals lined so that signals passed through them are as if they were a single thousands-mile wide crystal.

Even as recently as 10 years ago researchers had no idea at all that these features existed. They had pictured the planet as an onion with one smooth layer of rock after another—crust, mantle and core.

The new information explains a variety of phenomena, including why the plates of the earth's crust move and why such huge features as the Pacific Ocean exist, the paper said.

A Harvard University physicist, Mr Adam Dziewonski, said "We are beginning to get ourselves into a position where we can see and understand the whole machine inside the earth, at least in its basic parts d this is new."

Scientists now say that although the earth seems solid,its mantle ike a very slow-moving fluid over hundreds of millions of a per cent of the earth's mass is in its mantle that is the and is

Thesere speeded up, it would show an extremely hot hotter, less dens a flame on a stove, and above it the mantle from the surface sink The lighter elements rise to the top to form over several hundred mill On the earth, it is the crust that floats es".

The heaviest material, ironvection of the earth in which

Other features recently discovcrust. Colder, denser rock of material that used to be in the earthles back to the surface through its mantle.

The plates at the earth's surface collide with core. ciatingly slow force. At some edges or subduction zones of slabs forced under another and a great slab, cooler and denser thasinking below, sinks into the mantle.

The slab of the Pacific plate that is sliding under Asia is now believed to have flattened out at 400 miles down supporting the idea of an upper boundary. The sinking slabs remain cooler than the surrounding material for tens of millions of years.

All the continents were once bunched together in a grand mass called Pangaea. Since then the continents have broken up and drifted. The cold slab now under North America was ocean bottom in Pangaea days.

5

ATMOSPHERE, HYDROSPHERE, LITHOSPHERE AND THEIR CONSTITUENTS

Speaking in more general terms the earth, within the limits accessible for observation, consists of three important parts viz., the atmosphere (air and gases), the hydrosphere (water) and the lithosphere (the earth's crust).

1. ATMOSPHERE

It is the envelope of air which surrounds the earth. This envelope of air extends upto a considerable height from the surface of the earth. Since the atmosphere is not of the same density throughout and that atmospheric pressure decreases with height, it is a bit difficult to mark the outer limit of the atmosphere. There is, however, definite evidence that the atmosphere extends beyond 160 km. Although almost all of the atmosphere (about 97%) lies within 30 km of the earth's surface the upper limit of the atmosphere can be drawn approximately, at a height of 10,000 Km. The atmosphere is held to the earth by the gravitational pull of the latter. The atmosphere is densest at sea level and thins rapidly upwards. It constitutes a very insignificant percentage of the mass of the earth.

Composition of Atmosphere

From the earth's surface upwards to an altitude of about 80 km, the chemical composition of the atmosphere is highly uniform throughout in terms of the proportions of its component gases. In the uppermost reaches the atmosphere is charged with subatomic particles. Thus, except for the water vapour present, the composition of the atmosphere near the surface of the earth is practically uniform throughout the globe. Pure, dry air has very nearly the following constitution:

Nitrogen	78.03	percent by volume
Oxygen	20.99	" "
Argon	0.94	" "
Carbon-di-oxide	0.03	" "
Hydrogen	0.01	" "

All the component gases of the lower atmosphere are perfectly diffused among one another so as to give the pure, dry air a definite set of physical properties, just as if it were a single gas.

Structure of the Atmosphere

The structure of the atmosphere is highly complex but its layering is now well understood. The atmosphere has been divided into several layers according to temperatures and zones of temperature change. Altitudinally arranged the atmosphere falls into five layers or divisions such as:

(a) Troposphere
(b) Stratosphere
(c) Meso-sphere
(d) Ionosphere
(e) Exo-sphere

Troposphere

It is the lower most layer of the atmosphere. On an average, it extends up to a height of 12 km from the surface of the earth. At the equator, the thickness of the troposphere is the greatest i.e. about 18 km and about 8 kms thick over the poles. The troposphere contains about three-fourths of the total mass of the atmosphere, thus it is the densest of all layers. It is the locale of all the vital atmospheric processes which create the climatic and weather conditions on the earth's surface. The troposphere is characterised by:

(i) varying moisture content,

(ii) mobility of the air masses, both vertical and horizontal, and

(iii) regular temperature decline with height. The temperature of air in the troposphere decreases at the rate of 1°C per 165 meters of height.

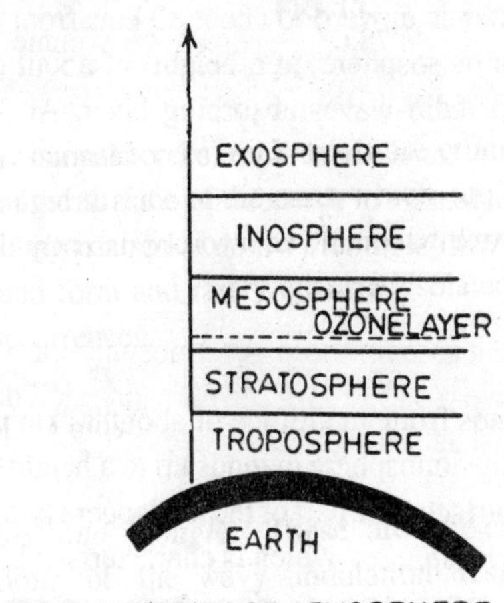

FIG.3. LAYERING OF ATMOSPHERE

Tropopause

It is an undefined region lying between troposphere and stratosphere. Here the temperature remains constant. The height of tropopause varies with latitude.

Stratosphere

It is the atmospheric zone extending from the upper boundary of the troposphere, to a height around 55 kms. In contrast to the troposphere, a steady rise in temperature with height is observed here. Here air is at rest. It is an isothermal region and is free of clouds, dust and water vapour. The upper strata is rich in ozone. The ozone layer serves as a shield, protecting the troposphere and earth's surface by absorbing most of the ultra-violet radiation found in the sun's rays, thus, is of great importance for the existence of life on the earth's surface. Since the water vapour content of the stratosphere is negligible, weather creating processes are never generated here. In the upper layers of the stratosphere the temperature rises to 0°C and higher. A stabilisation of temperature occurs at a height of around 55 kms and is called the STRATOPAUSE. Above it a fall in temperature is recorded.

Mesosphere

Above the stratopause lies the mesosphere, which is a very cold region. This layer extends upward to about 80 kms from the surface of the earth. Within the mesosphere, at a height of about 60 kms, there occurs a layer called radio-waves absorbing layer. At the end of the mesosphere, there is a transitional layer of minimum tempature of —80°C. An important feature of the mesosphere is its higher temperature in winter compared with summer, which is apparently due to a small ozone content.

Ionosphere

The ionosphere extends from an altitude of about 80 km upward. Studies have shown that the ionosphere extends up to a height of 1000-2000 km from the earth's surface. The part of the ionosphere lying between 80-800 km is called *thermosphere*, which is characterised by a steady rise in temperature.

In the ionosphere, almost all the atoms are ionised i.e this zone is made up primarily of ions (charged atoms). The ionosphere is much rarefied and therefore has a very low total mass despite a huge volume. This layer protects us from falling meteorites as it burns most of them. Besides, it reflects the radio-waves making wireless communication between places possible.

Exosphere

Above the ionosphere lies the exosphère. It is the outermost zone of the atmosphere. It is also known as the diffusion zone, where the atmospheric gases diffuse into the open space. It is prevented, in part, however, by the earth's magnetic field, which retains the ionised particles within the magnetosphere. Much about the exosphere is yet to be known.

2. HYDROSPHERE

It is a term used for the total body of water of the earth. In other words, all the natural waters occurring on or below the surface of the earth is known as hydrosphere. Thus the term includes the oceans, seas, lakes, rivers, snow and ice, underground and atmospheric water.

An enormous mass of water is concentrated in oceans and seas which occupy approximately 71 per cent (i.e 361,059,000 sq.km out of 510,000.000 sq.km.) of the surface of the entire globe. Oceans and seas communicate with each other and form a single mass of water called the *World Ocean*. Its average depth is about 3800 metres. The total volume of the world-ocean is about 1.4 billion cubic kilometres i.e about 97 percent of the world's free water is constituted by the world ocean.

The distribution of the oceans and seas is highly irregular in different latitudinal belts, as well as in the Northern and Southern Hemispheres. There are four major oceanic bodies:- (i) the Pacific, (ii) the Atlantic, (iii) the Indian and (iv) the Arctic. The first three oceans together constitute 90 per cent of the total area of water bodies. The Pacific is the largest ocean, both in surface area and volume.

Composition of Sea-Water

The sea waters are more variable in composition than the atmosphere, including large proportions of mineral matter as well as water and gases. Sea-water is a solution of salts. Dissolved salts, or solutes are added to the sea water from the erosion of the rocks of the earth's surface and from the eruption of volcanic materials espacially along oceanic ridges. These are lost by precipitation to ocean-floor sediments.

These salts result in the property of salinity, but the degree of salinity is not the same everywhere. The concentration of solutes in the water is affected by temporal and regional variations in erosion, precipitation and also by surface evaporation and the addition of water from rain and rivers. Thus, in the North Sea, for example, the percentage of salt is less than that of the Atlantic, in the Baltic it is very much less. In the Mediterranean, on the other hand, the proportion of salt is considerably greater than in any part of the open ocean.

The average salinity of the sea water is 35 parts per thousand i.e on the average 1000 grams of sea-water contains 35 grams of dissolved solids. According to Dittmar, the proportion of these solids are as follows:

Sodium chloride	—	27.213
Magnesium chloride	—	3.807
Magnesium sulphate	—	1.658
Calcium sulphate	—	1.260
Potassium sulphate	—	0.863
Calcium carbonate	—	0.123
Magnesium bromide	—	0.076
		35.000

The ingredients of sea water have maintained approximately fixed proportion over a considerable span of geologic time. Of the various elements combined in these salts, chlorine alone makes up 55 per cent by weight of all the dissolved matter and sodium makes up 31 per cent. Magnesium, calcium, sulphur and potassium are the other four major elements in these salts. Sea-water also holds in solution small amounts of all the gases of the atmosphere.

According to Brian Mason (Principles of Geochemistry, 1952) common elements present in the ocean-water are as follows

Elements		**Weight percentage**
Oxygen	—	85.79
Hydrogen	—	10.67
Chlorine	—	1.898
Sodium	—	1.056
Magnesium	—	0.127
Sulphur	—	0.088
Calcium	—	0.040
Potassium	—	0.038
Bromine	—	0.007
Carbon (inorganic)	—	0.003
Strontium	—	0.001
		99.718

The rest is made up by other dissolved gases. As we know, water absorbs oxygen more intensely than nitrogen. Different gases are differently absorbed by water. Thus while the oxygen to nitrogen ratio in the air is 1:4, in water these gases are usually found in a 1:2 ratio. Sea water derives oxygen from the air and also through photosynthesis by marine plants. The carbondioxide content of sea water is also high. Its sources are the atmosphere, river waters, the life activity of marine animals and volcanic-eruptions.

3. LITHOSPHERE

It is the general term for the entire solid earth realm i.e. crust. According to the recent concepts, the term lithosphere is used for the crust and the upper part of mantle, which is considered to be elastically very strong. This is the outer, cold part of the earth which is about 50-100 kms thick. Compared to the whole earth, the lithosphere is quite thin. Beneath the lithosphere, the rocks are still solid but are capable of creeping a few millimeters per year if the load on them is changed. The lithosphere is underlain by the asthenosphere which is considered to be a comparatively weaker zone.

According to the estimation made by Clarke and Washington, the lithosphere consists of 95% igneous rocks, 4% shale 0.75% sandstone, and 0.25% limestone (the metamorphic rocks being the altered equivalents of one or other of these rocks).

The average chemical composition of the lithosphere has been computed by a number of geo-scientists in terms of elements by weight percentage as indicated below:

Elements	Clarke and Washington (1924)	Brian Mason (1952)	A.B. Ronov A.A. Yaroshevsky (1976)
Oxygen	46.71	46.60	46.50
Silicon	27.69	27.72	25.70
Aluminium	8.07	8.13	7.65
Iron	5.05	5.00	6.24
Calcium	3.65	3.63	5.79

Sodium	2.75	2.83	1.81
Potassium	2.08	2.59	1.34
Magnesium	2.58	2.09	3.23
Hydrogen	0.14	—	—
Titanium	0.62	—	—
	99.34	98.59	98.26
Remaining Elements	0.66	1.41	1.74
	100.00	100.00	100.00

In terms of oxides (by weight) the chemical composition of the lithosphere is as follows

SiO_2	—	59.07
Al_2O_3	—	15.22
CaO	—	5.10
FeO	—	3.71
Na_2O	—	3.71
MgO	—	3.45
K_2O	—	3.11
Fe_2O_3	—	3.10
H_2O	—	1.30
TiO_2	—	1.03
Remaining Oxides	—	1.20
		100.00

The above tables indicate that 99 percent of the upper crust is made up mainly of 10 elements, with oxygen accounting for slightly less than 50%. Besides, the above mentioned ten oxides constitute more than 98 percent of the lithosphere, with silica being the most abundant one.

The mineralogical composition of the lithosphere has been computed as follows by Clarke and Washington in terms of volume percentage :

Quartz	—	11%
Alkali feldspar	—	16%
Plagioclase feldspar		
(Andesine)	—	47%
Amphiboles and Biotites	—	20%
Magnetite	—	5%
Apatite	—	1%
		100%

The above analysis represents only the average mineralogical composition of the lithosphere but it does not represent in any way the composition of the earth as a whole or even the crust as a whole.

6

THE ORIGIN OF ATMOSPHERE, OCEANS AND CONTINENTS

Most of the geo-scientists agree that the earth has originated from the cloud composed of gases and cosmic dust. Some of them believe that the protoplanetary cloud was hot while others think it was cool. Even though much remains unknown about the way in which the waters and gases accumulated around the earth, they both seem to have originated from the volcanic outgassing which has taken place since the formation of the planet.

On the basis of the assumed nature of the origin of the earth, two groups of hypotheses have been advanced to explain the origin of the earth's atmosphere and hydrosphere.

1. RESIDUAL ATMOSPHERE HYPOTHESIS

Advocates of this hypothesis believe that the earth was originally in a molten state and had a dense vapour saturated atmosphere, which later cooled first to a liquid and then slowly developed a solid crust. Thus according to this theory, the atmosphere and hydrosphere are resid materials from the primitive atmosphere that enveloped the eart the

Since hydrogen and helium are the most abundant elem

sun and because methane (CH_4) and ammonia (NH_3) are the most abundant gases in the atmosphere of the major planets of the solar system (excepting earth) which are also stable in the presence of an abundance of hydrogen and helium, it is believed that the primitive atmosphere was having hydrogen, helium, nitrogen and water vapour.

Helium being a very light and inert gas does not tend to form compounds to be retained by the force of the earth's gravitational attraction. With the continuation of cooling and chemical reactions, hydrogen compounds were formed which helped to retain hydrogen in the earth. Thus the gases were gradually transformed in to liquid and then into solid.

Objections

A.P. Vinogradov has indicated that if the molten planet had been initially surrounded by the heavy atmosphere, the water vapour present in such an atmosphere would have exerted a tremendous pressure. Under such conditions the rocks should have contained more water than it is being currently observed. Besides, the present day atmosphere should have preserved a large volume of inert cosmic gases (viz. Neon, Helium, Argon, Krypton, Xenon etc.) being the relict of the original atmosphere. On the contrary, there is a very small amount of inert gases in the present-day atmosphere.

2. ACCUMULATED ATMOSPHERE HYPOTHESIS

According to this theory, the atmosphere has accumulated from the degassing of the earth's interior and through chemical reactions that took place when the solid planetesimals accumulated. The cool terrestrial material was heated due to the adiabatic compression and decay of radioactive elements. Some would suggest that an original atmosphere, derived from cosmic gases, was largely lost as the earth heated up. The basis for this idea is that the atmosphere now has relatively small proportions of neon, argon, kryption and xenon compared with other planets. This event may have led to a new start in which most of the atmospheric constituents came from inside the earth: water vapour, carbon dioxide, carbon monoxide, nitrogen, chlorine, hydrogen and sulphur dioxide.

On cooling, the water vapour would condense in to the ocean. During a certain period of earth's history, liquid water was relatively sparse at the earth's surface. Volcanoes mostly started producing hydrospheric water and gave rise to small and shallow oceans which gradually reached its present dimensions. In this connection, it may also be indicated that the present ocean-floor is covered with numerous cones of extinct volcanoes. Thus, according to most of the researchers, the earth's atmosphere and hydrosphere could be produced by the volcanic outgassing.

Origin of the Continents

It has not yet been possible to ascertain the actual process by which the continents were formed. The views of various authors regarding the origin of earth are as follows:

(i) Because of the cooling and consolidation of the earth, on its surface wrinkles were developed. The upheaved portions formed the continents while the depressions where water got accumulated gave rise to oceans.

(ii) Some authors believe that unequal pressure of the atmosphere on the surface of the earth, when it was in a liquid state, is responsible for the formation of continents and oceans.

(iii) Chamberline, the proponent of Planetesimal hypothesis, believes the unequal accumulation of planetesimals on the surface of the earth gave rise to continents and oceans. Continents were formed where the accumulation was more and oceans where it was less.

(iv) Others believe that folding of the earth's surface caused the formation of continents and oceans. The anticlinal portion of the folds is considered to correspond to the continents and the synclines to the ocean basins.

(v) Some believe that continents began by the formation of island arcs and erosion of their volcanic materials exposed above sea-level to form sedimentary rocks. There is no record that this was the case, but the Precambrian-continents were probably smaller than those of today. Addition to continental areas takes place by the formation of island arcs offshore, or by the collision of an island arc with a continent.

The actual mode of origin of the earth is still continuing to be a controversial issue and so far not a single theory is able to confirm the actual mode of its formation.

7

GEOLOGICAL PROCESSES

It is an established fact that the development of the surface features of the globe is mainly due to the complex interaction of internal forces, atmospheric processes, rocks, ocean waters and living creatures. The surface of the earth is a zone where the rocks uplifted by internal earth forces come into contact with the atmosphere and hydrosphere. Thus, they are subjected to a range of processes powered by energy from the Sun. The processes which have been playing dominant roles in shaping the surface of the earth are both constructive as well as destructive in nature. All the geological processes can be conveniently grouped into two categories viz. endogenous and exogenous processes.

1. ENDOGENOUS PROCESSES

These are also known as hypogene processes. These are the processes of internal origin. In other words, a process which originates within the earth's crust is termed endogenous. These processes take place inside the globe and are governed by forces inherent in the earth and affected little by external influences. These processes cause phenomena, like earthquakes, emergence and development of continents, ocean troughs and mountain ridges, generation of volcanic activity, metamorphism of pre-existing rocks, deformation and movement of the earths crust both vertically and laterally etc.

The geomorphic features produced by these processes provide the setting for exogenous processes to operate upon. All features which owe their origin to an endogenous process are invariably modified by exogenous processes.

The endogenous processes are mostly caused by the thermal energy of the mantle and the crust. This thermal energy is derived from the decay and disintegration of the radioactive elements and from gravitational differentiation in the mantle. Some of the important endogenic processes and their role in the evolution of land forms are as described below :

(i) *Earthquakes* It is a form of energy of wave motion transmitted through the surface layers of the earth, ranging from a faint tremor to a wild motion capable of shaking builidings apart and causing gaping fissures to open up in the ground. The earthquakes are mostly produced due to underground dislocation of rocks.

(ii) *Tectonic movements* Tectonic movement of earth's crust are of various forms and are characterised by great complexity. In the course of geological history of the earth's crust, the rocks have been crumpled into folds, thrust over one another, broken up etc. giving rise to mountains, ridges, ocean trough and other landforms. The tectonic process of elevating or building up portions of the earth's surface is called diastrophism, which prevents the exogenous process from ultimately reducing the earth's land areas to sea-level. It is of two types viz. (a) Orogeny and (b) Epeirogeny. While 'Orogeny' refers to mountain-builiding activities with deformation of the earth's crust, *Epeirogeny* refers to regional uplift with marked deformation.

The lateral displacement of the crustal blocks are manifested in the phenomenon like continental drift, ocean floor. spreading etc.

(iii) *Volcanism* It is the phenomenon in which matter is transferred from the earth's interior and erupted onto its surface. It is one of the important manifestations of the dynamic nature of the earth. The process of effusion of magmatic material on to the surface of the earth, thus forming various volcanic structures and/ or flowing over the surface, is called volcanism.

Sometimes the magma on its way upward does not reach the surface and cools at various depths giving rise to magmatic bodies of irregular form, which are called *intrusives* or *plutons.* The phenomenon is known as *Intrusive magmatism* Even though the intrusions are not directly responsible for topographic features their existance in the upper crust of the earth may affect,to a great extent,the topographic features of the area formed by exogenous processes.

(iv) *Matamorphism* According to Turner & Verhoogen (Igneous and Metamorphic Petrology, New York, Mc Graw Hill Book Co. 1960) "the mineralogical and strutural adjustments of solid rocks to physical and chemical conditions which have been imposed at depths below the surface zones of weathering and cementations and which differ from the conditions under which the rocks in question originated" is known as metamorphism. Metamorphism involves the transformation of pre-existing rocks into new types by the action of temperature, hydrostatic as well as directed pressure and chemically active fluids. The main feature of the metamorphic processes is that the changes are isochemical and take place in solid state.

2. EXOGENOUS PROCESSES

These are the processes of external origin or, in other words, the processes derive their energy from sources external in relation to the earth viz. (i) energy from the sun which causes differential heating of the atmosphere giving rise to differences in pressures that make the wind to blow, sun's energy drives the hydrological cycle which involves the transfer of moisture from water bodies to atmosphere to land again to ocean etc. (ii) the force of gravity, (iii) the activity of organisms etc. Thus the exogenous processes are closely linked with the role of various external agents such as weathering, blowing wind, running water, underground - water, waves and currents in water bodies, (seas and oceans), glaciers etc. on the surface of the earth. Since these processes are restricted to the surface of the earth, they are called *epigene* processes. These processes constitute a very complex sum of mutually dependent changes i.e. all the exogeneous processes are involved with each other.

The exogenous processes act on the landforms to break up the rocks (weathering), to wear down the surface and carve out valley features (erosion) and the products of destruction are either dislocated under the influence of the force of gravity or are carried away by the blowing wind, flowing waters, moving glaciers etc. to lower areas like lakes, seas, oceans etc. where deposition takes place.

The term *Denudation* is used for the total action of all processes by which the exposed rocks of the continents are worn away and the resulting sediments are transported to suitable areas for deposition. Thus denudation is an overall lowering of land surface.

The exogenous processes tend to remove all the unevenness on the surface of the earth. As we know, the unevenness of the earth's surface is developed due to crustal movement, unequal erosion and deposition. The process by which the earth's surface irregularities are removed and a level surface is created, is known as *gradation*. All gradation-processes are directed by gravity. The processes of gradation are divisible into two major categories viz. *Degradation* and *Aggradation*.

Degradation is the process in which material from the high lands are removed by the geomorphic agents as a result of which the altitude of the highlands are reduced. Degradation of the earth's surface is mainly carried out through :

(a) Weathering,

(b) Mass-wasting, and

(c) Erosion.

Weathering is the process of mechanical disintegration and chemical decomposition of the rocks at the earth's surface, under the influence of factors like temperature fuctuations, water, oxygen, carbon-dioxide and organic life. It is a static process and does not involve any transport of the degraded rock materials.

Mass-wasting is the process that involves the spontaneous downward movement of soil, regolith and rock under the influence of gravity. Mass-wasting is usually aided by the presence of water but without any dynamic action of it or in other words the amount of water present is not sufficient to act as a transporting medium. The process occurs at speeds ranging from those which are so slow that the movement is imperceptible to rapid flow and catastrophic slumping and rockfalls. Evidences of the down-slope movement of rock and soil is found almost universally.

Erosion is the process associated with the geomorphic agents like wind, river, glaciers etc. by which rock materials are loosened or dissolved and then transported from one place to the other. According to Arthur Holmes (Principles of Physical Geology, Second Edition, The English Language Book Society and Nelson, 1975) "all the destructive processes due to the effects of the transporting agents are described as erosion".

Erosion consist of processes like:

(i) collecting together of the loose material produced by weathering.

(ii) wearing down the surface and carve out valley features which is commonly known as abrasion or corrasion.

(iii) mechanical wear and tear of the transported materials while they are in transit by the geomorphic agents, and

(iv) chemical solution through the dissolving power of the geomorphic agents like river-water, sub-surface water etc, which is also known as *Corrosion*.

Aggradation is the process of deposition of sediments. As we know, under favourable conditions, when the transporting agents lose their carrying power, the transported materials get deposited, sometimes in the sea, sometimes on the land. Thus the low lying tracts are gradually filled up through deposition of sediments by running water, ground water, wind, glaciers, wave, currents, tides in seas, oceans etc.

Thus, in the nature the process of gradation is considered as a three-fold process because the earth's surface is first decayed and eroded, secondly the products of the decay and erosion are transported and finally the transported materials are deposited in low lying areas.

The geological processes, as already indicated, play significant roles in shaping the surface of the earth. The details of the role played by each individual geomorphic agent are discussed in separate chapters of this book.

8

EARTHQUAKES

An earthquake is a sudden and temporary vibration set up on the earth's surface, ranging from a faint tremor to a wild motion, due to the sudden release of energy stored in the rocks beneath the earths surface. Earthquake is a form of energy of wave motion which originates in a limited region and then spreads out in all directions from the source of disturbance.

Earthquakes usually last for a few seconds to a minute. Sometimes, the vibrations are so feeble that we can not feel them, whereas the violent earthquakes result in huge material loss and the loss of human lives.

The point within the earth where earthquake-waves originate is called the *focus* and from the focus the vibrations spread in all directions. They reach the surface first at the point immediately above the focus and this point is called the *epicentre*. It is at the epicentre where the shock of the earthquake is first experienced. It is, however, evident that no earthquake can possibly originate at a mere point alone. Earthquakes occur beneath the surface of the earth, where the rocks yield suddenly, of course, after porolonged build-up of stresses. They are often associated with fault-lines, which provide a zone of fracture and easy yielding. Earthquake emerges at various depths, which may be anywhere in the crust or as far as 700 km down into the mantle.

Due to the sudden yielding of rocks to stresses, waves of energy are sent out through the earth. These waves of energy are called *seismic-waves*. The seismic waves emerge in the focus of an earthquake and radiate outward, like ripples produced when a stone is thrown into a pool of water, gradually losing energy.

TYPES OF SEISMIC WAVES

It is known that the seismic waves pass through the earth is mainly because of its being an elastic body- i.e. when a stress is applied it becomes deformed more and more, but when the stress is removed it returns to its original shape. If the stress applied is too great the elastic body will yield permanently. The seismic wave have been considered to be the outcome of the elastic deformation of rocks. Since, all elastic bodies can be subjected to two types of deformation, compression and shear, the seismic waves are also related to these.

On the basis of their amplitude, wave length and nature of vibration the seismic waves are classified into three main types namely-*P-wave*, *S-wave*, and *L-wave*.

(a) *Primary or P-waves* These are longitudinal or compressional waves, similar to the sound waves. In this case, rocks vibrate parallel to the direction of wave propagation i.e. in the same direction as the waves are moving. This kind of wave motion leads to the longitudinal compression and rarefaction of the substance. These waves travel in all the media i.e. in solid, liquid as well as gaseous, even though their propagation is inconsistent as far as the velocity is concerned.

Some authors consider them to be true sound waves because they are often audible as deep rumblings or even as loud reports in the early stages of earthquakes. In general the P-waves travel at 1.7 times the speed of the S-waves. They have short wave length and high frequency. Due to their superior speed the P-wave are the first waves to arrive at points distant from the place of earthquake origin and are therefore called *Primary-waves*. These are also known as *Push waves*, Longitudinal waves etc. The velocity of P-wave depends on density and compressibility of the medium.

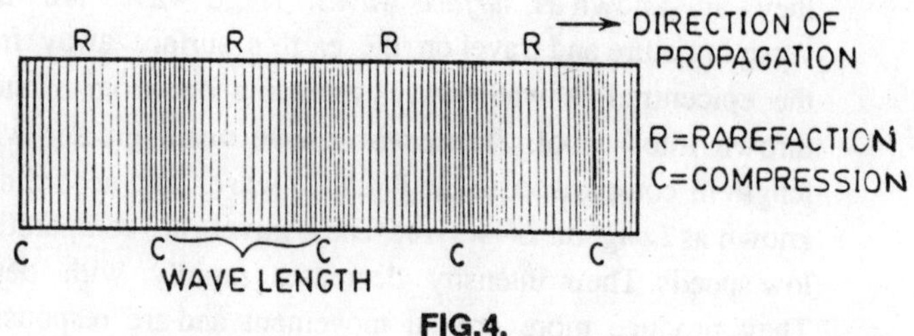

FIG.4.

(b) *Secondary or S—waves* These are transverse waves and move like the oscillations in a piece of string vibrated sideways from one end. In this case, therefore the rocks vibrate perpendicular to the direction of wave propagation. These waves are also known as shear-waves as they are capable of changing the shape of the material without changing its volume. As the liquid and gaseous substances do not exert resistance to the change in form, the shear waves are not propagated through these substances. However, these waves travel in solid media. These waves move less rapidly than the "P-waves". Besides, their velocity varies through the solid parts, proportional to the density of the materials. They are also having short wave-length and high-frequency. Since transverse vibrations cause a shaking of the earth's surface, these are also known as *shaking-waves.*

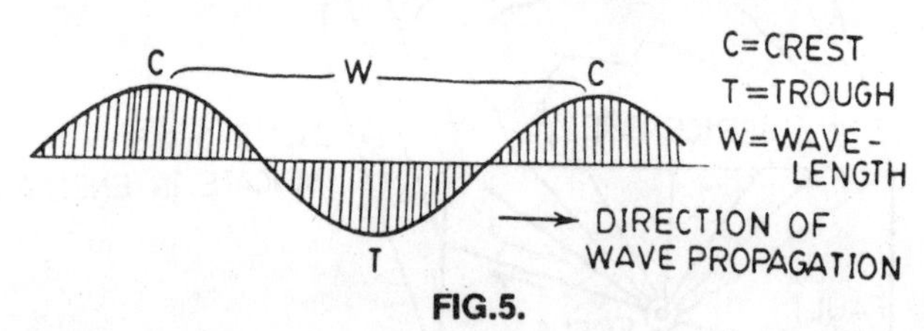

FIG.5.

Because the *P* and *S—waves* travel through the earth's interior and spread outward from the focus in all directions they are known as *Body-waves.*

(c) *Surface or L-waves* They are also known as Rayleigh waves or Love waves or Long waves. These waves are generated by the energy brought to the surface by the P and S-waves. These are confined to the outer skin of the crust. Therefore, they are known as *surface waves.* These waves are transverse in nature and travel on the earth's surface away from the epicentre like water waves spreading out from a stone thrown into pool. These waves have much greater wave length in comparison to the *P* and *S waves* and accordingly known as *Long* or *L—waves*. They travel at comparativly low speeds. Their intensity decreases rapidly with depth. They produce more ground movement and are responsible for most of the destructive force of the earthquake. Thus, these waves cause great damage to the lives and properties, during an earthquake.

In a particular earthquake-record the lines joining the places where the shock arrives at the same time are called *homoseismals* or *Coseismals* or *Homoseists* . In a similar way, a line joining all places experiencing the same earthquake intensity is known as an 'Isoseismal-line or in other words, it is an isodiastrophic line of equal damage.

The diagram given below indicates the isoseismal lines and their relation to the epicentre and to the wave paths radiating from the focus of an earthquake.

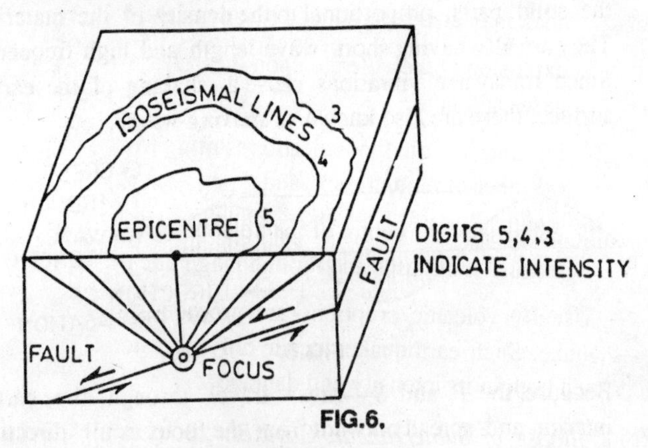

FIG.6.

CAUSES OF EARTHQUAKES

Earthquakes originate due to various reasons which fall into two major categories viz. non-tectonic and tectonic.

(1) *Non-Tectonic Causes* The non-tectonic causes of earthquake in

clude those associated with the geological agents operating upon the surface of the earth, volcanic eruptions as well as with the collapse of subterranean cavities.

(a) *Surface causes* Perceptible vibrations may set up by—

(i) the dashing waves and crashing breakers along the sea-shores;

(ii) abrupt descending of running water from a higher altitude as in the case of water-falls in adjacent areas;

(iii) rock falls and avalanches in mountains, large landslides etc.

Apart from the above, there are also artificial surface-causes which produce perceptible tremors, for example, underground explosion of bombs, passage of trains and tanks, working of heavy machinery in industrial areas, explosion in mines, failure of dams under the pressure of the impounding water etc.

(b) *Volcanic-causes* Volcanic earthquakes occur around active volcanoes mainly due to explosive eruption and also due to the hydraulic shocks of magma that forcibly fills underground chambers and channels. A shock may also be produced by any of the following reasons:

(i) explosion of the volcano upon the release and expansion of gases and lava,

(ii) faulting within the volcano resulting from pressures in the chamber of molten rock, and

(iii) collapse of the centre of the volcano into the space formed by extrusion of gases and molten magmatic materials.

Usually volcanic eruptions are preceded or accompanied by earthquake. Such earthquakes occur only occasionally and are more localised both in its intensity and damage.

(c) *Collapse of Subterranean Cavities*: Sometimes because of the removal of support from below, by the action of underground water, the ground surface subsides or collapses suddenly producing local tremors. This is usually noticed in the caverns of Karst areas.

(2) *Tectonic Causes* About 95 per cent of all the earthquakes are due to sudden earth movements along existing or new faults. The association

of earthquakes with fault-lines is an established fact. As such, earthquakes caused by faulting or folding in the crust are known as tectonic earthquakes. The term *tectonic* (Greek word tekton means a builder) refers to the structural changes of the crust due to deformation or displacement. Such earthquakes generally result from sudden yielding to strain produced on rocks by accumulating stresses.

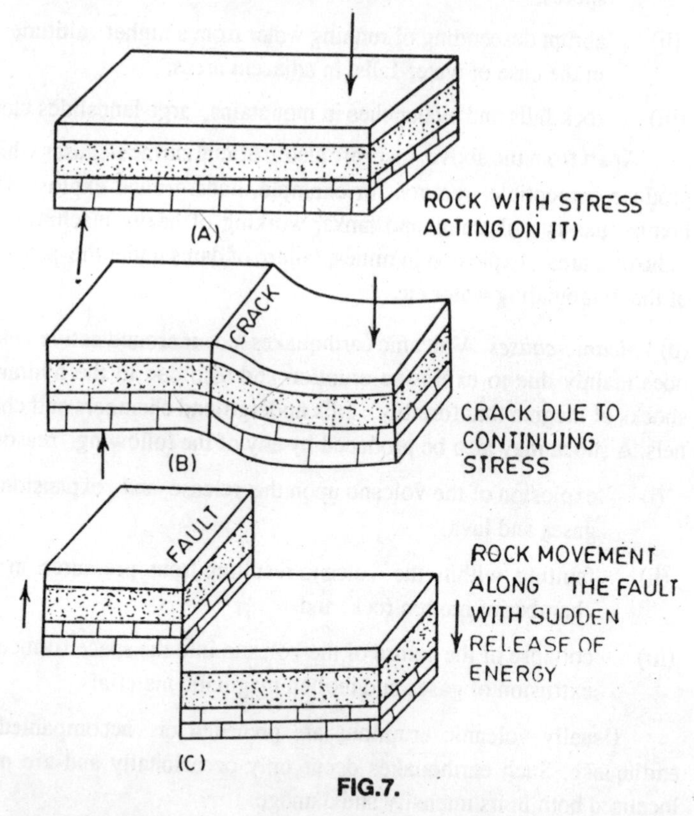

FIG.7.

Prof. H.F. Reid has proposed the *Elastic Rebound Theory* to explain the origin of tectonic earthquakes. According to this theory materials of the earth, being elastic, can withstand a certain amount of stress without undergoing a permanent deformation. When the stress exceeds the elastic limit a crack or fracture is developed. Frictional resistance along the fracture prevents the fractured blocks from being slipped off from each other and thus promotes a build up of strain. With the superimposition of more stress the rock units on either side of the fracture are subjected to enormous strain and there comes a stage when the rocks can not bear more straining i.e. at the time when the frictional resistance is overcome. At this stage, there occurs a sudden slipoff of the fractured blocks to position of no strain. Thus the energy stored in the system through decades is released instanteneously causing underground dislocation of rocks and waves of energy are sentout through the earth. These waves of energy are called seismic waves that set up vibrations on the earth's surface. The crack or fracture alongwhich the displacement of rocks occurs is known as a fault. Thus, according to the "Elastic Rebound Theory", earthquakes are commonly associated with movement along a fault.

The preceding diagram, indicates the elastic rebound theory. Most of the earthquakes probably take place due to tectonic reasons, which is further supported by the fact that the regions of intense and frequent earthquakes coincide with those characterised by new intense tectonic movements.

CLASSIFICATION OF EARTHQUAKES

Earthquakes are usually classified on the following bases:

(a) Cause of origin;

(b) Depth of focus; and

(c) Intensity and magnitude of earthquake.

(a) *Cause of Origin*: On the basis of the causes of earthquake, they are classified as:- (i) Tectonic and (ii) Non-tectonic earthquakes. The non-tectonic earthquakes are mainly of three types due to surface causes, volcanic causes and collapse of cavity roofs.

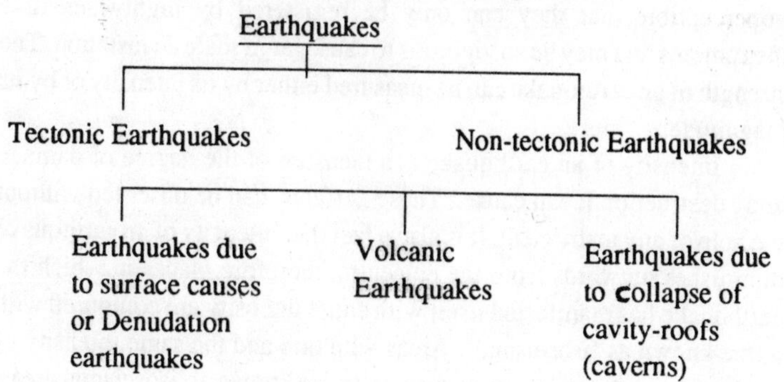

(b) *Depth of focus* As we know, the instrument designed to detect seismic waves is called seismometer and the seismograph is a seismometer to record the earth vibration. This record of earth vibration is known as seismogram. It has now become possible to estimate the depth of focus of an earthquake by analyzing seismograms. On the basis of the depth of focus, earthquakes are classified as:

(i) Surface-earthquakes

(ii) Shallow-focus earthquakes or normal earthquakes.

(iii) Intermediate-focus earthquakes, and

(iv) Deep-focus earthquakes.

Surface-earthquakes are those in which the depth of the focus is less than 10,000 metres. The earthquakes with the hypocentre at a depth of 10 to 50 kms are known as *shallow-focus earthquakes.* When the earthquake is originated at a depth of 50 to 300 Kms, it is called *intermediate-earthquake.* The deep-focus earthquakes or the plutonic earthquakes are those with hypocentres located at depths more than 300 kms. Majority of the deep focus earthquakes originate between 500 and 700 kms. Shallow-focus earthquakes constitute about 85 percent of all the earthquakes and the intermediate and deep-focus earthquakes account for 12 and 3 percent respectively of all the earthquakes. Thus it is seen that the intermediate and deep-focus earthquakes together account for only 15 percent of the earthquakes.

(c) *Intensity and Magnitude of Earthquakes*

As we know, the tremors caused by earthquake may be so feeble and imperceptible that they can only be registered by highly sensitive instruments and may be so vigorous to cause large scale devastation. The strength of an earthquake can be measured either by its intensity or by its magnitude.

Intensity of an earthquake is a measure of the degree of damage and destruction it can cause. These effects can be observed without the help of any instrument. It is also a fact that intensity of an earthquake diminishes outwards from the epicentre, therefore places in which the earthquake has manifested itself with equal intensity are contoured with a line known as 'isoseismal'. Areas with one and the same intensity of earthquake restricted by isoseismal lines are known as isoseismal areas.

There is a number of disadvantages in using intensity as a measure of the strength of a particular earthquake, some of the important disadvantages are as follows:

(i) The strength of an earthquake decreases with distance from its epicentre. Thus different degree of damage occurs at different distances for the same earthquake.

(ii) The degree of damage depends much on the geological characteristics of a particular area as well as the type of construction, population—density etc.

However, two scales of intensities are in vogue viz. (i) Rossi Forrel's Scale and (ii) Mercalli Scale.

Rossi-Forrel's Scale

According to this scale, there exists ten distinct and well-defined intensities beginning from *I* and ending with *X* of which *I* represents the most mild earthquake while *X* the most disastrous one. The scale is as follows:

Intensity Number	Name	Effects
I.	Imperceptible	Recorded by sensitive instruments only.
II.	Feeble	Recorded by all seismographs and felt by experienced persons only.
III.	Very Slight	Felt by several persons at rest. It is strong enough for the duration and direction to be recorded.
IV.	Slight	Felt by persons in motion. Moveable objects disturbed. Affects window doors, ceilings etc.
V.	Weak	General alarm, ringing of bells, disturbance of furniture and beds etc.
VI.	Moderate	General awakening of persons from sleep, stopping of clocks, visible oscillation of trees etc.

VII.	Strong	Overthrows moveable objects, general panic, fall of plaster from the walls without damage to the building etc.
VIII.	Very Strong	Fall of chimneys and cracks in the walls of buildings.
IX.	Severe	Partial or total destruction of buildings.
X.	Destructive	General destruction of buildings rock-falls and landslides in mountaineous regions.

Mercalli Scale

This scale is developed by Mercalli, an Italian seismologist, after he made studies of the intensity and regional effects of earthquakes. The scale had at first ten divisions but later on it was modified to a scale of 12 degrees. The higher the number of intensity the greater is the damage. The scale is as follows:—

Intensity	Name of the shock	Effects produced
I.	Instrumental	Hardly noticeable. Detected only by instruments.
II.	Very Feeble	Felt in some cases by people at rest. Delicately suspended objects may swing.
III.	Feeble	Felt quite noticeably, like passing of a truck. Perceived by persons at rest.
IV.	Moderate	Felt by people in motion, windows and doors may start vibrating standing automobiles may rock noticeably.
V.	Relatively Strong	Most sleepers wake up, plaster peels off in places, unstable objects move, bells ring.

VI.	Strong	Felt generally; people get panicky and run outdoors, cracks develop in the walls.
VII.	Very Strong	Slight damage to buildings, there are fissures in the walls and chimneys are broken.
VIII.	Destructive	Buildings are impaired, heavy furniture overturned, chimneys fall, persons driving motor car disturbed.
IX.	Ruinous	Buildings collapse, ground cracks conspicuously, underground pipes broken
X.	Disastrous	Buildings are seriously damaged or destroyed, land-slides on steep slopes, rails bent, ground badly cracked.
XI.	Very Disastrous	Few building remain standing, bridges destroyed, numerous fissures appear on the earth's surface, great landslides and floods.
XII.	Catastrophic	Total destruction; waterfalls and lakes emerge; waves seen on ground surfaces; objects thrown into air; river channels change their course.

Magnitude of Earthquakes

As we know, an earthquake invariably involves the sudden release of a certain amount of energy stored in the rocks beneath the earth's surface. The potential energy is transformed into kinetic energy in an earthquake, which is propagated into all directions from the focus in the form of elastic seismic waves. Magnitude is a measure of the energy released during the earthquake. This is determined on the basis of amplitude of seismic waves recorded as seismogram.

Professor Charles Richter of the California Institute of Technology proposed a scale of earthquake magnitude in 1935 to indicate the quantity of energy released by a single earthquake. This is a numerical scale of magnitudes from 0 to 9, with higher numbers indicating larger earthquakes. This scale is commonly called the Richter-scale. Richter has numbered the scale in steps, with each step representing an earthquake record ten times larger than the previous step, for example an earthquake of magnitude '8' is ten times larger than an earthquake of magnitude '7' and hundred times larger than that of magnitude '6' and thousand times larger than the magnitude of '5.' Since the Richter scale is logarthmic, the difference between two consecutive whole numbers on the scale means an increase of ten times in the amplitude of the earth's vibrations. The largest earthquake so far measured is 8.6. Earthquakes up to '6' on the Richter scale do not cause serious damages.

LOCATION OF EPICENTRE

There are two principal methods to locate the epicentre of an earthquake viz. (i) Plotting of isoseismal and (ii) Determining the position of the epicentre through mathematical means using the seismograms of various recording stations for a particular earthquake.

Ploting of Isoseismals

As it has already been indicated isoseismals are lines joing points of equal intensity of an earthquake. The isoseismal lines encircle one another and the isoseismal for the highest intensity lies just around the epicentre. In this method, the location of the epicentre can be made approximately.

Mathematical Method

Here in this method, the first step is to determine the 'epicentral distance' i.e. the distance of the epicentre from the recording station. The mathematical formula used for calculating the distance of the epicentre, is as follows:—

$$D = T \frac{V_p V_s}{V_p V_s}$$

where D = Distance of the epicentre from the recording station i.e. the epicentral distance.

T = Time interval between the arrival of primary and secondary waves, as recorded in the seismogram i.e. *tp*—*ts* where

tp and ts are the times of arrival respectively of P and S waves at the station.

Vp and Vs = These are the velocities of the P and S-waves respectively. Vp is always greater than Vs since the path of the waves are the same.

$$\frac{D}{Vs} - \frac{D}{Vp} = T \text{ or } D\left(\frac{1}{Vs} - \frac{1}{Vp}\right) = T$$

$$\text{or } D\left(\frac{Vp - Vs}{Vs\ Vp}\right) = T \text{ or } D = T\left(\frac{Vs\ Vp}{Vp\text{-}Vs}\right)$$

It is possible to determine D' because T, Vp and Vs are known.

For a particular earthquake once the epicentral distance is determined for any particular station, the same method is followed for at least three different stations and arcs with radii proportional to the calculated epicentral distances are described geometrically as shown in the diagram given below

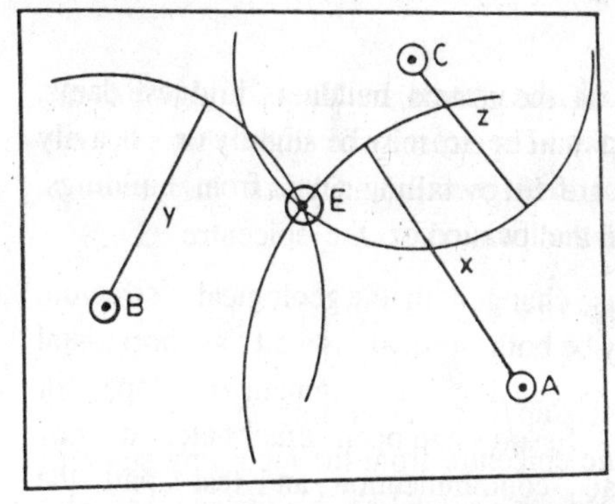

FIG.8.

The point of intersection of the three arcs represent the epicentre of the earthquake.In case the arcs do not meet at a point, the area enclosed by them should be the epicentral tract of the earthquake.

Effects of Earthquakes

Earthquake is a natural calamity of such a type that it never gives opportunity and scope to people to save their lives and escape. People are caught unaware and the catastrophe is so sudden and the frequency so uncertain that there is not much scope to caution people before hand.

There are earthquakes almost every year causing large-scale damage and devastation. On the basis of the statistics, it has been estimated that earthquakes take an average yearly toll of 14000 lives and cause damage to extensive property. But, it should be borne in mind that all the earthquakes are not of equal strength and that the extent of damage depends on the degree of acceleration with which the ground rocks during the earthquake.

As we know, earthquakes are due to the sudden release of energy stored in the rocks beneath the earth's surface. The seismic waves that result from such release of energy cause the main or major shock, besides there are occasional minor disturbances that take place subsequent to the main or major shocks called the *after shocks.* The after-shocks prevail for a long period and can cause sufficient damage to the structures already weakened by the major shocks.

Damage by earthquakes varies with the strength of the earthquake, local bedrock, type of building construction and life-support systems. Some of the important effects of the earthquakes can be summarised as follows :

(i) Due to the vibration of the ground, buildings, bridges, dams, poles and posts and fences etc. may be slightly or heavily damaged and people are hit by falling debris from buildings. Railways are buckled and twisted.

(ii) Earthquakes also cause changes in the geological structure of an area. There may be both vertical as well as horizontal displacement of rocks causing development of slopes or scraps and sometimes fissures and open cracks etc. It may also destroy the road communication, and tear apart the water pipes and gas pipes etc.

(iii) Landslides and subsidence of land also take place during an earthquake. Sudden subsidence of the land near sea or lake cause flooding and drownings. In the Peruvian earthquake of 1970, million tons of ice, snow, rock and boulders moving at a tremendous speed (estimated at 480 km per hour) buried the town of Yungay and all its inhabitants.

(iv) Ground water and its movement gets disturbed by earthquakes; besides, the courses of streams and rivers change, new springs develop and in certain favourable conditions sand-dykes may also develop.

(v) Fire is a usual problem associated with earthquakes, due to broken gas and water mains and fallen electrical wires.

(vi) *Tsunamis* An important secondary effect of a major earthquake is the seismic sea wave, or tsunamı as is known to the Japanese. An earthquake below the sea-floor generates seismic sea-waves which often have catastrophic consequences. They usually devastate the costal regions.

They are caused usually by earthquakes below the sea-floor, submarine landslides or volcanic explosions. The diagram given below shows the generation of a tsunami:

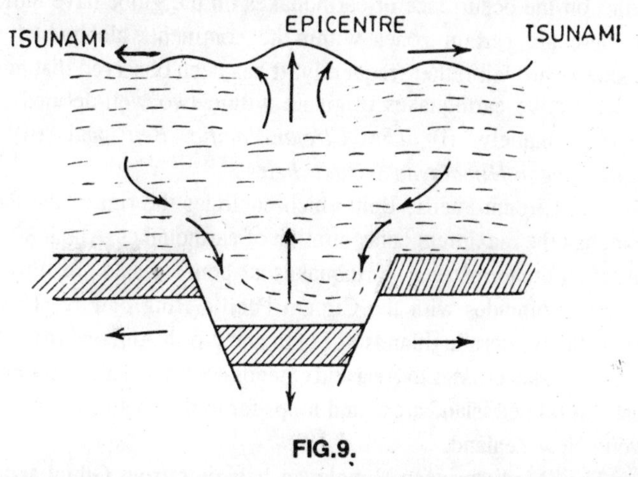

FIG.9.

During an earthquake, sometimes there is a sudden subsidence or upheaval of the sea floor, due to displacement of blocks. Accordingly, all the water at the epicentre of the earthquake is lifted or dropped for an instant giving rise to a sea-wave of several hundred kilometers long but only a few metres high in the open sea. This sea-wave moves at a speed of 750 to 800 kms per hour. Even though the speed of the wave slows down drastically as it moves through shallow coastal water, the height of the wave rises to 30 or 40 metres on approaching the coast. This wave of water, as a very large and fast wave, hits the shore but because of its extremely long-wave length it does not withdraw quickly as the normal waves do. Its long duration and great height cause great damage to the entire coast and many deaths by drowning in low-lying coastal areas. It is thought that coastal flooding which occured in Japan in 1703, with an estimated life loss of 1,00,000 persons, may have been caused by seismic sea waves. In the year 1952, a strong earthquake with the epicentre at the northern tip of Kuriles trough gave rise to a Tsunami that reached the shores of Kamchatka and Kuriles and caused enormous destruction,

GEOGRAPHIC DISTRIBUTION OF EARTHQUAKES

Earthquakes are distributed unevenly on the globe. In certain places they are more frequent and intense whereas in other places they are extremely rare and feeble or their effect is hardly perceptible. Studies on the occurrence of earthquakes on the globe have indicated that there are certain zones within the continents alongwhich seismic shocks are felt rather frequently. It has been observed that most of the destructive earthquakes originate within two well-defined zones or belts namely (i) *The Circum-Pacific Belt* and (ii) *The Mediterranean-Himalayan Seismic belt.*

The Circum-Pacific Belt which encircles the rim of the Pacific ocean, has the maximum concentration of earthquakes. About 80 percent of all the terresterial earthquakes is concentrated in this belt. This ring coincides with the Circum Pacific Ring of Fire. This belt follows the western highlands of South and North America from Cape Horn to Alaska, crosses to Asia and extends southward along the eastern coast and related island arcs, and loops far to the south east and south beyond New Zealand.

The Mediterranean-Himalayan belt runs from Gibraltar to the East via the Atlas mountains, the Pyrenees, the Apennines, the Balkan mountains, the mountain chains of Asia Minor, the Caucasus, Hindukush, the Himalayas, the mountain chains of Burma and the islands of Indonesia, where it meets the Circum-Pacific belt in the north of Australia.

Apart from the above two belts, a number of shallow-focus earthquakes also occur in the zones of mid-oceanic ridges as well as in the body of the volcano or in its peripheral parts during eruption.

It is noticed that the present earthquake regions are associated with the younger fold-mountain regions and the present earthquake activity is a phase of the end of the Alpine-Orogeny.

PREDICTION OF EARTHQUAKES

Even though there has been tremendous progress in the field of science, in the recent years, it has not yet become possible to have any scientific method for accurate prediction of earthquakes. Attempts to say in advance when exactly the earthquake will start have met with little success. However, several techniques are being developed for scientifically forecasting a coming earthquake.

In the prediction of an earthquake, three main questions are involved as follows:

(i) To indicate the place where underground shocks would occur ;

(ii) To indicate the anticipated strength of the future earthquake;

(iii) To indicate the time of its occurrence.

The indication about the place and the strength of the coming earthquakes can be provided to a satisfactory extent by maps of seismic zones. It involves the study and analysis of the frequency of the previous earthquakes, the location of their epicentres and hypocentres etc. However, in regions that have not been affected by earthquakes the problem becomes complicated. In such cases, a close study of the geological structure of the seismic areas is made against the background of the geological structure of the region in which earthquakes have never been registered. In case the geological features are found to be identical, then the seismic regime will be approximatly the same. But in the absence of the identical features, the region should be regarded as aseismic.

By using the map of seismic zoning it becomes possible to anticipate the strength of a future earthquake, for a particular region.

So far as the question (iii) as stated above is concerned, it is difficult to answer satisfactorily when an earthquake will strike. The basic difficulty for prediction seems to be due to abrupt and sporadic occurrences of earthquakes. Majority of the methods used for forecasting a future earthquake involve monitoring slight changes that occur in rock next to a fault before the rock breaks or moves. Some of the important methods used for prediction of earthquakes are as follows:

1. Soviet geo-scientists used earthquake waves from other, unrelated earthquakes to measure the wave speed through the

volume of rock around a fault. As we know, while an earthquake is in the making, stresses gradually built up within a given area of the earth's crust with attendant changes in the elastic properties of rocks. Therefore , the velocity of propagation of earthquake waves that pass through these rocks also changes. It is possible to measure the changes of wave velocity and register the moment when this velocity reaches a certain critical value, at which an earthquake is likely to occur. American seismologists have also confirmed the utility of this method.

2. In some regions, the surface of the earth tilts particularly in the epicentral zone before an earthquake occurs. This is based on the fact that in areas of tectonic movements, the surface of the earth continually experiences small deformations, tilting and bending. Through the use of highly sensitive instruments like tilt indicators which register the slightest flexure and movement of strata, it is possible to establish the critical value of deformations at which mechanical strain must be relieved i.e. at which an earthquake should be expected.

3. Foreshocks can also give certain indications about the future earthquakes. These foreshocks can be detected, with the help of seismographs, microphones and other sensitive instruments, which are otherwise practically imperceptible shocks and noises that accompany the making-process of an earthquake.

4. Changes in the electrical resistivity of the rock, rock magnetism, porosity, water pressure and other physical properties of rocks may indicate an imminent quake.

5. Variation in well-water composition (particularly the radon content) and its temperature takes place several days before an earthquake. Besides, because porosity changes are related to strain and since it affects the water-level in wells, the changes in water-level in wells take place before an earthquake.

6. Earthquakes can also be predicted by watching animal behaviour. For example, animals of burrowing habitat leave their burrows before an earthquake. Birds start flying in panic, horses become skittish and suddenly animal behaviour be-

comes abnormal before an earthquake.

Efforts are still in progress to find out a more scientifically reliable method for the prediction of earthquakes. Some of the new findings are as follows:

(A) According to a news-item published very recently in the Hindustan Times (25 the November 1987), Earthquakes are related to the sun's rhythm. The published item is as follows:

Earthquakes to the Sun's Rhythm

The occurrence of earthquakes follows the sun's eleven-year cycle, according to Oleg Barsukov of the Soviet Institute of Earth Physics, says APN.

The earthquakes tend to occur in greater numbers during the years when the sun-spot activity is maximum or minimum, says the report. This is based on the instrumental observations of earthquakes by the world seismic network, for the period 1931 to 1982.

Out of the 6,200 earthquakes with magnitude exceeding six recorded during this period 111 events a year occurred during the years of high solar activity and 126 events per year during the sun's quiet years.

(B) Immediately after the occurrence of the Bihar-Nepal earthquake, the predictions of Mr. N.K. Agarwal published in the Times of India (Aug, 23, 1988) reads as follows:

Engineer Predicts More Quakes

The earthquake that shook Bihar, West Bengal and other parts of north India yesterday, killing thousands and injuring hundreds, was part of a series of natural calamities predicted by an Indian civil engineer in April this year.

Mr. N.K. Agarwal, the civil engineer who had predicted that starting from May or June this year, earthquakes and other natural calamities would rock different parts of the country up to June 1989, said a number of leading scientists all over the world had now come out in support of his predictions.

He claimed that as predicted by him, earthquakes measuring 6.5 on the Richter scale occurred twice in the month of June this year, confirming the prophesy of the 16th century astrloger Nostradamus. It was just luck that the tremors did not involve substantial loss of life and property.

Mr. Agarwal said researches at the U.S. National Oceanic and Atmospheric Administration (NOAA) had indicated that the sun's activity was increasing at the fastest rate since NOAA started recording its observations in IRP.

Mr Agarwal pointed out that another U.S. scientist, Dr Jim Shirely of California, had also said that the present increase in solar activity could be matched by only two periods earlier during the past 13 centuries from 1623 to 1633 A.D. and from 1810 to 1812 A.D. On both occasions intense volcanic eruptions had occurred.

Mr Agarwal said Dr Shirely's prediction that the present solar cycle would usher in similar periods of volcanic and climatic extremes on earth was in agreement with his own predictions reported in April.

He endorsed predictions of Dr Shirely and another expert of the U.S. Geological Survey about major earthquakes in the next year of two. Dr Shirley has predicted a steep rise in solar activity over the next 18 months, reaching levels higher than anything seen in this century.

Mr Agarwal said that the quake was second in a series of major earthquakes that would continue to rock vast parts of India, Nepal and Pakistan in the coming months. On August 6 earthquakes of the intensity of more than 6.5 on the Richter scale had rocked north, north-eastern and eastern India. The August 6 quake was the most intense witnessed in India since the devastating Assam earthquake of 1950.

He felt that other countries like the U.S. China, Japan and Chile falling in the seismic belt, would witness more serious volcanic eruptions and quakes during the coming months.

(C) Mr. Arun Bapat has suggested a new method called Seismic grid method to correctly predict an earthquake. His prediction and the procedure of his method as published in the Times of India (24th Aug. 1988) after the occurrence of Bihar earthquake is as follows:

Quakes May Rock 3 More Countries

PUNE, August 23: Severe earthquakes are likely to hit Indonesia, the Philippines and Venezuela in the near future, according to Mr. Arun Bapat, a Pune seismologist who had correctly predicted the earthquake which hit north Bihar and eastern Nepal.

Mr Bapat said the magnitude of the earthquakes would be

between 7 and 7.5 on the Richter scale. There would be heavy loss of life and property as these countries are thickly populated.

Accoring to him, other places which would be hit by earthquakes are Mexico, California, and the Alaska peninsula. They would be of almost similar intensity and might take place in about a year.

He said the advanced countries had better facilities of monitoring and, therefore, could take precautions to avoid damage to a considerable extent.

Mr Bapat, who is the head of the earthquake engineering division in the Central Water and Power Research Station (CWPRS) here, which comes under the Ministry of water resources, had predicted the earthquake that rocked Assam and its surroundings on August 6. He had also given correctly the magnitude range. However, the earthquake occurred about 70 km. away from the place he had predicted.

Mr Bapat's predictions are based on a method. known as seismic grid method developed by him. In this, the seismological parameters are divided in a particular fashion and the available seismic data is examined.

The area is divided into grids of ten degrees latitude by ten degrees longitude. The earthquake magnitude is divided into four categories: (a) between 1 and 4.9, (b) five to 5.9,(c) six to 6.9 and (d) seven to above. The time of observation selected is one decade.

The seismic grid method consists of observing the number of earthquakes per grid, per magnitude range and per decade. These observations are supported by a seismic model which involves various stages of activity.

While making the prediction, the first stage is called establishment stage in which the seismic gap is established as a result of two major magnitude seismic events.

The next stage is the development stage, during which a few medium intensity earthquakes occur. The next stage is the maturity stage of the gap' during which small-magnitude earth quakes occur around the gap. Once a seismic gap reaches maturity level, the area could be positively identified as a highly vulnerably seismic region which may experience an earthquake in the near future, that is within 500 to 1,000 days.

The seismic grid area, which is about 10,000 sq km, could be examined in terms of geo-physical, seismological and meteorological parameters and the epicentre of occurrance reasonably foretold. It is somewhat difficult to tell the exact time of occurrance, according to Mr. Bapat.

Using this method, it was possible to predict the Mexico earthquake of September 19,1985. This method has been described in a paper jointly written by Mr Bapat and Mr. R.C. Kulkarni and presented at the eighth world conference on earthquake engineering held at San Francisco in July 1984.

Mr Bapat said the work done by distinguished seismologists in India. China, the U.S., the Soviet Union, Japan and other seismically active countries had raised hopes that prediction of earthquakes would be possible within the foreceable future.

Though Mr Bapat has given detailed accounts of his theory and the scientific data on which it is based, the scientific community as well as the government has not accepted it. However, he continues to publicise his findings.

(D) The opinion of Dr. Haresh Saha, about the new methods of predictions published as a news item on 29th of Aug, 1988 in the Times of India is as follows.

Quake in Himalayas can shake up Delhi

A massive earthquake in the Himalayas can shake up buildings taller than 20 stories in distant Delhi, a leading seismoligist of Stanford University has said.

"Just because your city is far away from a potential seismic spot, you cannot ignore safety in designing tall buildings" Dr. Haresh Shah warned.

The long distance effect resulting from *special amplification* is something that scientists learned from the Mexico earthquake about three years ago, according to Dr Shah who is on a visit to the structural engineering research centre in Madras.

The quake that originated far away from Mexico avoided several nearby cities but only destroyed the Mexico City built on alluvial soil, he said.

The soil filtered some, but amplified other frequencies close to the natural frequencies of the buildings that collapsed because of the phenomenon called *resonance*. Dr Shah said.

Delhi, like several cities on the Gangetic plains is also located on alluvial soil that can magnify the effect of a quake occurring far away in the Himalayas, the source of major earthquake in India.

Given a choice, structures like atomic plants should be built on nonfractured rock and not alluvium, according to Dr Shah.India's fourth atomic station at Narora on the banks of the Ganges is sitting right on alluvium.

Elaborating on earthquake research, Dr Shah said despite the $500 million being spent by the US gelological survey, scientists have no idea how to predict earthquakes. '*Forecast is in the realm of the future*,' he said.

Those who claim they can predict are either *charlatans or gypsies*, according to Dr. Shah, who expressed surprise over reports in Indian newspapers about scientists having predicted the recent Bihar earthquake.

A prediction that does not give the size of the earthquake, the exact place and the day of occurrence is of no value to anybody, Dr. Shah said, adding that the government should be cautious about taking actions on the basis of predictions made by individuals.

A prediction few years ago led to eviction of panic stricken people from Pisa in Italy and when the quake did not arrive the Italian government had a politcal crisis on their hands, Dr Shah said.

Describing the procedure in the U.S. Dr Shah said that "anytime a qualified scientists comes up with a prediction, the National Science Academy sets up a panel to verify the claims. It is then the panel that informs the state governor. More often the predictions proved wrong", he said.

According to Dr Shah, the only prediction that had so far proved right was the Heinchang earthquake in China a few years ago when two million people were evacuated to safety.

EARTHQUAKE CONTROL

In the light of the *elastic rebound theory*, it has been discovered that release of strain could be made by injecting fluid to lubricate fault planes. Instead of letting strain build up to be released subsequently in a major destructive earthquake, it may be possible by fluid injection to reduce the frictional resistance to faulting. With the fluid reducing the friction, strain can be released in a number of smaller and timed shocks. In such cases. the injected fluid does not cause the quake but it allows a quake that would occur sometime anyway to be controlled.

Earthquake-Proof-Construction

Since the damage caused by an earthquake depends on the rate of vibration of the quake, the construction should be made accordingly.

It has been seen that if the structure is made quite firm, the shocks may not be able to cause much damage to it and the structure can withstand the vibrations. In such buildings the roofs are made as light as possible. However, in regions where earthquakes are frequent, the structures should be made of lighter materials.

SOME IMPORTANT EARTHQUAKES IN INDIA

1. Cutch Earthquake of June, 16th, 1819
2. Bengal and Kashmir Earthquake of 1885
3. Assam Earthquake of 1897, 1935, 1950, 1988 (6th Aug 1988)
4. Kangra Earthquake 1905, 1975, 1987
5. Bihar Earthquake 15th Jan 1934, 21st Aug 1988
6. Koyna Earthquake 11th December 1967.

RECENT EARTHQUAKES IN INDIA

Recent Earthquakes in India-North Eastern Region

It was on 6th of Aug 1988, several parts of north, north eastern and eastern India were rocked by moderate to severe earthquake causing extensive damage to buildings and disrupting rail and tele-com link.

Two shock waves with their origin in the Hindukush-region and Indo-Burma border rumbled across Assam, Meghalaya, Arunachal Pradesh, Nagaland, West Bengal, Mizoram, Jammu & Kashmir, Punjab and New Delhi, forcing people out of their homes. Panic-stricken people rushed out of their houses as the quake accompained by rumblings ran through rattling doors and windows. The earthquake epicentre in the Hindukush region as recorded by the seismograph at Delhi at 2.06 p.m. (IST) indicated to be at a distance of about 1000 km. The seismological observatory at Shillong reported that the quake had its epicentre 295 km south east of Shillong. The Coloba observatory in Bombay registered two earthquakes at 6.11 am and 2.37 p.m. with epicentres between 1923 km and 1971 km north-east of Bombay.

Intensity The quakes measured 6.5 to 7 on the open ended Richter-scale.

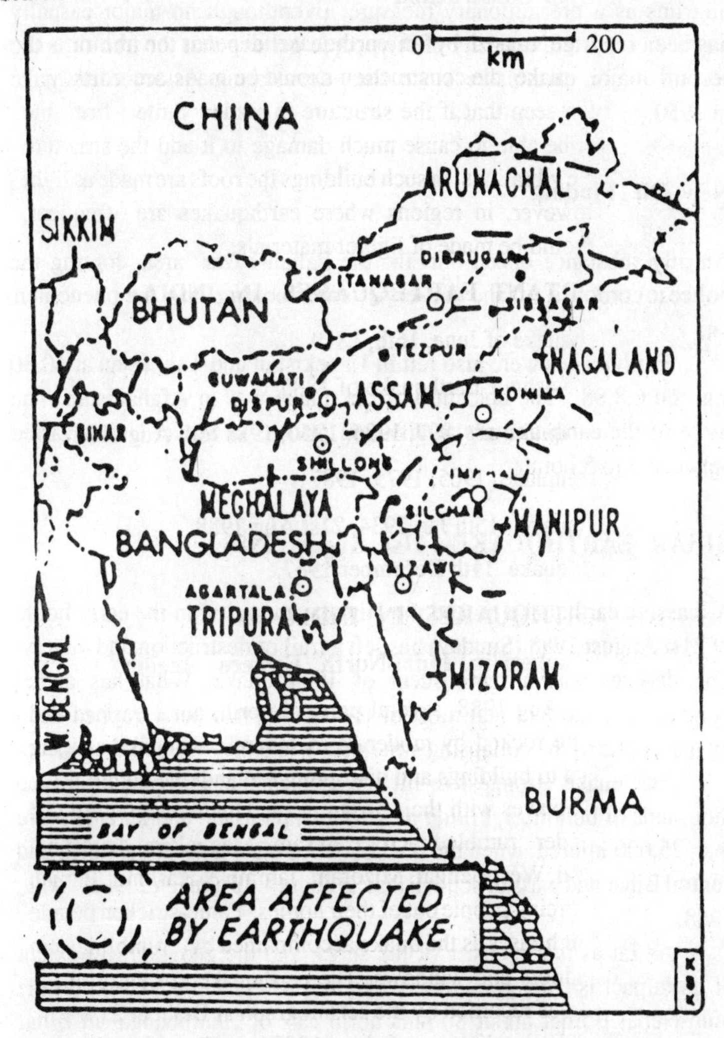

FIG. 10. Area-Affected by Earthquake in N.E. (India)

Damage A hotel building caved in by a metre on central road in Silchar during the quake, normal life, power supply, rail services were greatly disturbed. In Shivsager, bricks, concrete lumps of a Shiva temple gave away prompting the district authorities to prohibit entry of pilgrims as a precautionary measure, Eventhough no major casualty has been reported, majority of the people believe that the tremor is the second major earthquake since the devastating Assam earthquake in 1950.

New Oil Finding

An oily substance oozed out in the Paltan-bazar area forcing the police to cordon it off and oil-experts are looking into the phenomenon.

Earthquakes were also felt in Uzbekistan and Tajikastan at 10.30 a.m. on 6.8.88. The epicentre of the quake was in Afghanistan. The force of the earth-tremors at different places in Soviet Union varied between 3 to 5 points.

BIHAR EARTHQUAKE—21st August 1988

A massive earthquake rocked the Himalayan region in the early hours of 21st August 1988 (Sunday) has left a trail of destruction and misery. The disaster was a reminiscent of 1934 quake. What has added poignancy is the fact that most of the victims had been trapped dead by the debris of the collapsing structures while they were in their sleep.

The quake strongest to hit the area in the 54 years destroyed thousands of buildings, killing more than a thousand and leaving more than 25,000 injured, when the quake rocked the whole of the north and central Bihar and parts of south Bihar at about 4.40 a.m. on 21st of Aug, 1988.

As far as the intensity of the shock and the geographical extent of its impact is concerned, the quake had its epicentre located on the Indo-Nepal Border about 80 kms north east of Dharbhanga in Bihar recorded 6.7 on the open ended Richter scale, although China's State Seismological Burueau is stated to have cited it to be 7.1. That Uttar pradesh, Madhya Pradesh, West Bengal and all the north-eastern states, besides Bangladesh, also experienced tremors testify to the severity of the visitation. The worst hit districts were Dharbhanga, Madhubani and Monghyr though its impact was also felt in several other districts including Begusarai and Saharasa.

The one minute quake flattened houses, breached flood protection dykes and triggered landslides. While the initial tremor lasted one minute, several after shocks were felt until 6 a.m.

In several parts of the State, rail and road communications were disrupted following collapse or sinking of tracks, damage to the bridges and roads. The Kosi river was rising alarmingly.

Earthquake in Nepal

Above 700 people have been killed and more than thousands injured when the quake rocked the Himalayan kingdom of Nepal at 4.54 a.m. local time (4.40 a.m. IST) on 21st August 1988. Huge landslips caused by the quake have blocked traffic in the Dharan Dhankuta Highway in Eastern Nepal. A 1 km stretch of the Mahendra Highway has got elevated by two feet starting from Lahan, also in eastern Nepal. Most of the wells in the area got filled up with sand.

FIG.11 The building of Land Development Bank at Darbhanga was razed to the ground in last Sunday's earthquake on 21st August 1988.

The Nepalese Mines and Geology department said that the quake recorded 5.7 on the Richter-Scale and lasted 40 to 60 seconds. The epicentre of the tremor was located at 26.7° latitude and and 86.6° longitude around Udayapur in Sagarmatha zone of eastern Nepal, 170 Km South east of Kathamandu.

This is a worst disaster to occur in the region after 54 years. In the cataclysm that hit the area on 15th January 1934, as many as 10,700 human lives were lost and several towns in Bihar were wiped out of existence by a quake that touched 8.4 in the Richter scale.

The present Bihar Nepal quake (21/8/1988) is being described here as the second in magnitude and effect to the great earthquake of 1934.

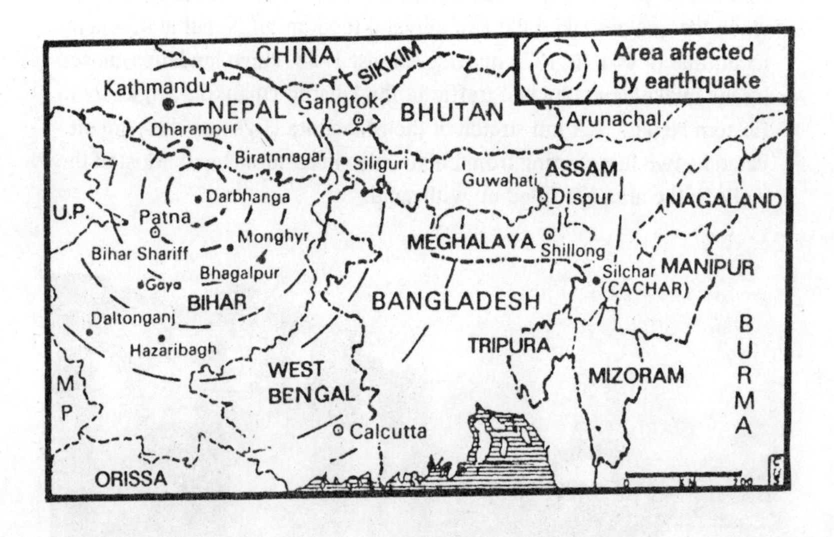

FIG. 12. Area affected by the Present Earthquake (21st August 1988).

CONCLUSION

As we know, there have been more disastrous earthquakes before. The ones in Kansu (China) and Tokyo claimed 1,80,000 and 1,43,000 lives in 1920 and 1923 respectively. The worst quake took place in

Calcutta on 11, October,1737. The toll then was over 3,00,000 and the world's highest in earthquake. More recently, earthquakes claimed 67,000 lives in Peru (1970), more than 700,000 in Tangshan (China) in 1976 and more than 20,000 in Chile in 1980. The 1985 earthquakes in Mexico claimed about 7,000 lives. In March 1987, more than 4,000 died in a series of quakes in Ecuador.

Earthquakes in the 20th century have killed nearly two million people and damaged property worth Rs 10,000 crores. The last major earthquake in India was in Assam on the Independence Day in 1950 killing 1530 persons.

The recent devastating earthquake in the Bihar Nepal border was not quite unexpected. Northern India falls in the active seismic zone stretching from Indonesia and Burma to Paksitan, Afghanistan, West Asia and the Mediterranean. In the instant case, the epicentres lay in what seismologists call zone VI which denote the most seismological active regions of the country. In contrast almost the entire peninsular India falls in Zone 'O', Which is least seismologically active.

As is well known, quakes result from the shifting of vast sections of the earth's crust. Although the movement is only a few inches, the energy released by it produces shock waves or tremors.

Not long ago, Indian, Italian and Soviet scientists carried out an important geophysical experiment in the Pamirs, the Hindukush and the Himalayas in order to study the abyssal structures under these mountains. Another important objective was to verify an old hypothesis explaining the congestion of mountain ranges in the region by the continuing northward advance of the Hindustan platform.

According to this hypothesis, the giant continental plate *dives* under the Asian continent pushing it up, thereby causing tremendous fissures, giant cracks in the earth's crust. It is for this reason, experts argue that most earthquakes in India are due to the movement of the Indian plate against the Eurasian plate by *Continental drift*.

As a result of this push, enormous forces are generated which build up in rocks over months and years until they become large enough to cause the rock strata deep down to give way releasing the pent up energy as a big quake.

As this process goes on all along the Himalayan belt, extending from the Karakoram and Hindukush ranges in the west to the Arakan Range in the east, the entire region is earthquake prone. The entire region falls in zone IV and Zone V considered to be seismologically highly active.

M. Sami Ahmed, K. Z. Amani and Bashid Umar from the Dept. of Geology, Aligarh Muslim University, Aligarh are of some other opinion about the causes of Bihar earthquake. Their opinions are as follows: "The cause of the Bihar Nepal earthquake can be attributed to an array of factors. But in our opinion, it was a water induced earthquake. The epicentre of this 6.7 intensity earthquake on the Richter scale entailing large scale destruction of lives and property in the region was at Dhran in Nepal where 99 per cent of all the houses collapsed. The huge amount of water in the barrage at Dharn must have injected itself forcefully into the east-west regionally extensive geofractures that characterise this part of the Himalayas. The water on encountering high temperature zones down below must have been converted into vapour with enormous pressure leading to the movement of the rocky blocks and thereby generating the earthquake waves.

The focus of this shallow earthquake most probably lies below Dhran. Moreover, it is probable that the large quantity of water in the barrage near Dhran must have been attracted by the gravitational pull of the moon and might have aided the process.

It is dangerous to construct huge reservoirs in seismically active zones like the Himalayas (viz., Tehri and other dams). One can well imagine the dimensions of destruction entailing the structural failure of the dams due to high intensity earthquakes. This should act as an eye opener for the engineers and planners of this subcontinent."

Despite claims of success in predicting earthquakes, precise prediction of the time and area of an earthquake, enabling people to take precautions, remains a distant proposition. Given this scenario, strengthening of defences in order that its potential for damage is minimised should get the top priority, and in the art of withstanding quake shocks Japan seems to be far ahead of other countries. And this means evolving and applying innovative construction technology, zoning regulation and building code, besides communication and training system attuned to disaster management. The fact that very little damage was done to Tokyo in 1985 by an earthquake measuring 6.2 points on the Richter scale, shows what innovating modern building technology, zoning regulations and building codes can achieve.

9

TECTONIC MOVEMENT

Tectonic-movements belong to the class of Endogenous processes which take place inside the globe and are governed by forces inherent in the earth. The process that involves the breaking and bending of the earth's crust under internal earth forces is known as *tectonic-movement.* The science dealing with the structures of the upper layer of the earth, i.e. the crust and the upper mantle, and the movements producing these structures is called *Geotectonics.*

It is an established fact that rocks do not always behave in the way as they do at the surface. Rocks, igneous, sedimentary and metamorphic, are more or less solid and in the surface condition they show elastic behaviour, viz. compression (i.e. shortening in dimensions due to compressive stress), elongation (i.e. stretching or increase in the dimension of the body along the line of action of tensional stress), shearing (under tangential or lateral stress), as well bending, twisting etc. But rocks within the earth's crust experience greater confining pressure as a result of the load of overlying rocks etc. and the depth of burial, high temperature due to the general geothermal-gradient and also the presence of gases and solutions coming from magmas. As such they do not behave strictly as elastic material but as *elalstico-viscous materials,* a term coined by E.W. Spencer for the substances exhibiting the characteristics of an elastic solids under some conditions and those of a fluid under other conditions. Apart from the gravitational stress, a rockmass within the earth's crust, is also acted upon by a number of stresses setup by processes like igneous intrusion, flow movements in the mantle, tidal stresses (Caused by the influence of other bodies in space,on the earth) contraction and expansion of some parts of the crust etc. These forces cause the rocks to crumple into folds, thrust over one another or to break-up etc. so that the landforms on the surface also undergo changes, mountains are formed and deep basins are developed. All these crustal deformations mostly occur as a result of the differential stress acting upon the rocks. As a whole the crustal defor mations are known as *Diastrophism.*

Sedimentary beds are the most suitable ones for preserving the records of crustal disturbances because of the fact that these beds exhi bit an undisturbed and almost horizontal dispostion over a vast area of the earth's crust and even a slight bending,breaking or some sort of distortion of the strata is easily detected in them. Thus the sedimentary rocks serve as an index of the nature of the movements which have occurred.

As a direct result of diastrophism, irrespective of the size or scale, four major groups of structures are produced, which are as follows:

1. Folds
2. Faults.
3. Joints.
4. Unconformities.

Originally the sedimentary rocks are horizontal, but owing to movements of the earth's crust, they are often tilted out of their original position. Sometimes the tilting of the beds even takes the originally horizontal beds to a vertical position, but more often the beds, due to unequal uplift or subsidence of the crust (in any particular region) moves into an inclined position. The attitude of an inclined bed is defined by two elements, namely *Dip* and *Strike*.

Dip is essentially the angle of inclination of a bed with respect to a horizontal plane. The dip of a bed has got two components like direction as well as magnitude. The angle of dip varies between 0° (For horizontal beds) to 90° (For vertical beds). The direction of dip is the geographical direction, along which a bed has maximum slope.

Strike is the line of intersection of a bed with the horizontal plane.

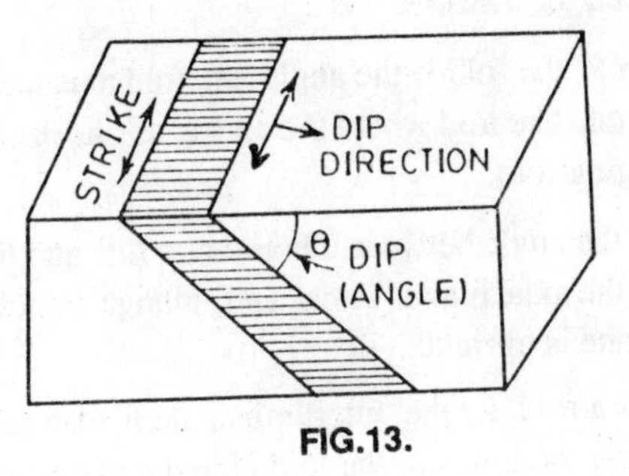

FIG.13.

Folds

Due to compressional forces, rock-strata are crumpled, forming wavy undulations on the surface of the earth which are known as folds. The process of folding is also known as *plicated deformations*. Folds are of various size and form and rarely occur as isolated feature. They have widespread occurrences.

Elements of Folds

(i) *Crests and troughs* These are the convex and concave portions of the wavy undulation respectively. Thus the wavy undulations are formed of a series of alternate crests and troughs.

(ii) Core is the inner part of a fold.

(iii) Limb is the stretch of the rock-beds lying between any crest and its adjacent trough i,e, the sides of the fold. It is also known as flank.

(iv) Axial plane is an imaginary plane which divides the fold as symmetrically as possible. It is, therefore, that any point on the axial plane is at equidistance from both the limbs. The axial plane may be vertical, inclined or horizontal or even a curved surface.

(v) Axis is the line of intersection of the axial plane and the ground surface.

(vi) Hinge is a line running through the points of maximum curvature of any of the beds forming the fold. It is also known as fold bend or flexure.

(vii) Plunge of the fold is the angle the fold axis makes with the horizontal. The fold where the axis is not horizontal is known as plunging fold.

(viii) Pitch is the angle between a horizontal line and the azis measured on the axial plane. Pitch and plunge coincide when the axial plane is upright

(ix) Angle of a fold i,e the interlimb angle is formed by the lines continuing the limbs of the fold upto the place of intersection.

(x) Height of the fold is the vertical distance between the hinges of the adjacent anticline and syncline. It is also known as the amplitude of a fold.

(xi) Width of a fold is the distance between the axial lines of two adjacent anticlines or synclines.

(xii) Nose or curve of the fold is the part of the fold adjacent to the hinge line. It is also known as hinge-area

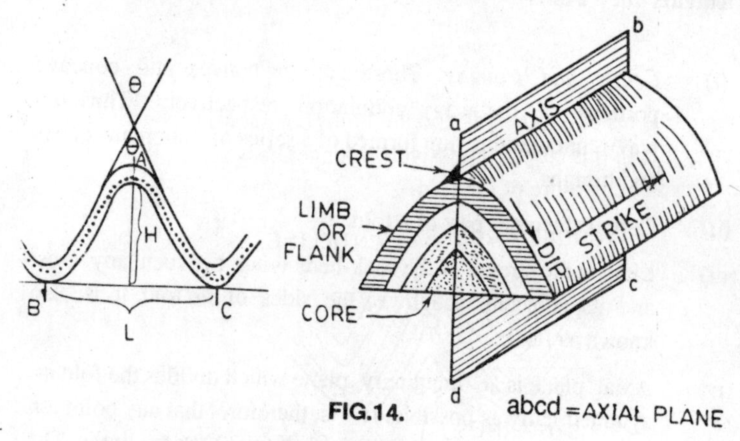

FIG.14. abcd = AXIAL PLANE

L = Width of the fold (BC)
H = Height of the fold
A = Axis of the fold
θ = Angle of the fold

AC & *AB* = Limbs of the fold
B & *C* = Troughs

Types of Folds

Several types of folds have been recognised on different basis as follows:

(a) *Appearance in Cross-section* Folds like antiform, synform, anticlinorium, synclinorium, anticlinal bend (monocline), synclinal bend (structural terrace), anticline and syncline are included in this category (The details of these folds are beyond the scope of this book). In fact all these folds are simple or complex modificatin of anticlines and synclines

Anticline

It is a folded structure where the strata are convexed upwards and the limbs commonly slope away from the axial plane. Here, while relatively younger beds are found upwards, the older rocks constitute the core i.e. the centre of curvature

Syncline

This fold is concave upwards and the limbs commonly dip towards the axial plane. In this case progressively younger beds are found towards the centre of curvature of the fold.

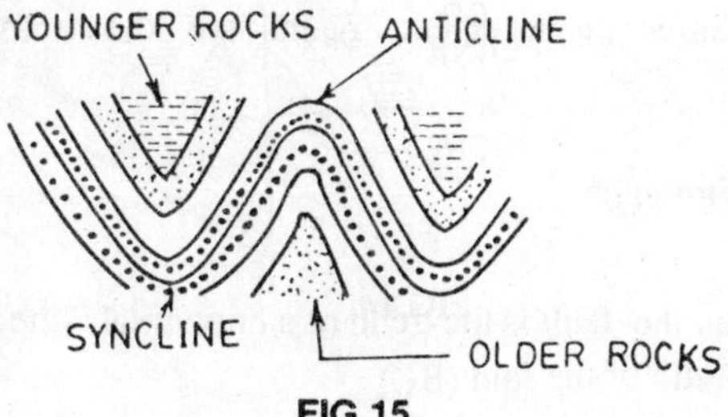

FIG.15.

(b) *Symmetry of fold* On this basis folds have been classified as symmetrical, asymmetrical, recumbent, isoclinal, overturned and homoclinal folds. In these folds, the axial plane may be either vertical, inclined or horizontal.

(c) *Thickness of limb* Three important forms of fold have been identified on this basis viz. parallel fold, similar fold and suprateneous fold.

(d) *Inter-limb angle* Folds like open or gentle fold, closed fold, tight fold and cylindrical fold etc. are recognised on the basis of the interlimb angle while the inter-limb angle is more than 70° in the case of open folds it ranges between 30° to 70° in closed and is below 30° in case of tight folds

(e) *Attitude of folds* On this basis folds like plunging, non plunging doubly-plunging, perclinal folds and reclined folds have been recognised.

(f) *Mechanism of folding* According to the mechanism of folding the following four types of folds have been identified, viz. Drag-fold, Flexure-fold. Shear-fold and Flow-folds.

(g) *Mode of origin* On the basis of the mode of origin folds may be classified as tectonic and non-tectonic folds. While generative folds, culminations and depressions are of tectonic origin, diapiric folds, cambering and valley bulging etc. are folds of non-tectonic origin.

Apart from the above types of folds, there are special types of folds like chevron folds, box or coffer folds, fan-folds, kink-bands, geanticlines etc also.

FAULTS

These are well-defined cracks alongwhich the rock-masses on either side have relative displacements Faults are known as disjunctive dislocations.

Fault Terminologies

(i) Strike of the fault is the trend of a horizontal line in the plane of the fault.

(ii) Dip of the fault is the angle between a horizontal surface and the plane of the fault. It is measured on a vertical plane that strikes at right angles to the fault.

(iii) Hade is the complementary angle of dip i.e. the angle which the fault plane makes with the vertical plane. In other words, hade = 90°— Dip ·

(iv) Fault plane is the plane along which the displacement takes place.

(v) Fault scrap is an upstanding structure with a steep side which is formed due to the relative displacement on either side of the fault line.

(vi) Hanging wall and foot wall. In case of an inclined fault, one of the dislocated block apears to rest on the other. The former is called the hanging wall and the latter the foot wall. In a more simple

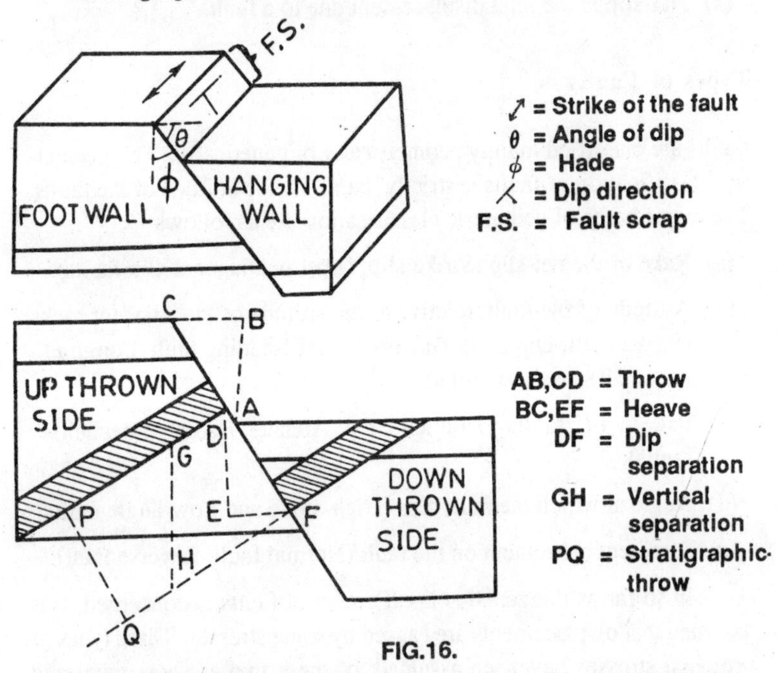

FIG.16.

way, it can be stated that the fault covers the footwall and lies beneath the hanging wall or in other words, the part above the fault plane is known as the hanging-block and the one below it is the foot wall.

(vii) *Up-thrown and down-thrown side* Of the two blocks lying on either side of the fault plane, one appears to have been shifted downwards in comparison with the other the former is therefore known as the down thrown side and the latter 'upthrown side'.

(viii) *Throw and heave* The throw of a fault is the vertical component of the apparent displacement of a bed measured along the direction of dip of the fault. Similarly, the horizontal component of the apparent displacement is known as heave or gape .

(ix) *Stratigraphic throw* If the same bed occurs twice because of faulting, the perpendicular distance between them measured along a vertical section, at right angles to the strike of the fault is known as stratigraphic throw.

(x) Net-slip is the total displacement due to a fault.

Types of Faults

Faults are classified mainly geometrically or genetically. The geometric classification of faults is strictly based on the attitude of the faults. The major bases of geometric classification are as follows

(a) Rake of the net-slip (Strike-slip, Dipslip, Diagonal-slip faults).

(b) Attitude of the fault, relative to the attitude of the adjacent rocks (Strike fault, Dip fault, Diagonal fault, Bedding fault, Longitudinal and Transverse faults)

(c) Pattern of faults (Parallel, Step, Arcuate, Radial, Enechelon faults).

(d) Angle at which the fault dips (High-angle and Low-angle faults).

(e) Apparent movement on the fault (Normal fault, Reverse fault).

In so far as the genetic classification of faults is concerned, it is assumed that displacements are caused by some stresses. Three types of principal stresses havebeen assumed, of these two are horizontal and the third one is vertical and due to gravity alone. The variety of faults depends on the orientation of the three principal stresss, three sets of conditions in which all the stresses are compressional may arise; accordingly three types of faults originate viz. Normal faults, Thrust faults and Transcurrent fault.

Normal Faults

In these faults, the maximum stress is vertical, and both the mean as well as the minimum stress is horizontal. In such cases, the hanging wall has moved relatively downwards and the fault plane dips toward the downthrown side when the plane of the fault is vertical, the fault is called vertical-fault. The normal faults are also known as *gravity* or 'tensional faults. To this category belong the Horsts, Grabens, Dip-slip faults, Antithetic and Synthetic faults, Parallel, Bedding, Step-faults, etc.

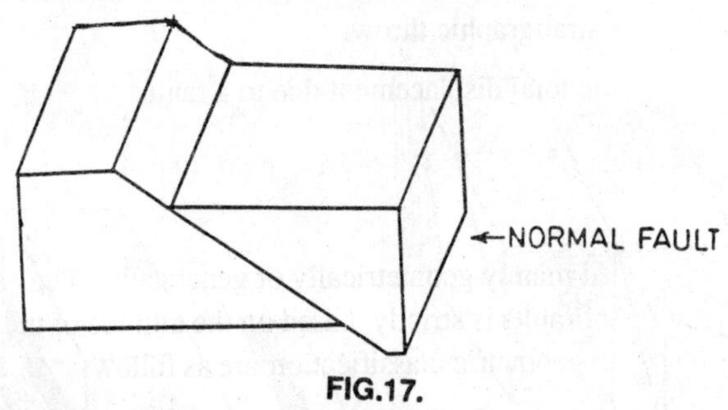

FIG.17.

Thrust Faults

In these faults, the maximum as well as the mean stress are horizontal and the minimum stress is vertical. In such cases, the hanging wall moves relatively over the footwall. where the dip of such faults is more than 45°, they are known as Reverse faults. In other words, the low angle reverse faults are known as *thrust-faults*. This type of fault is regarded as a shear deformation under the conditions of compression of the earth's crust.Nappes, Imbricate or Schuppen structures, etc. belong to this category of faults.

Strike Slip Faults

These are the faults, where the maximum and minimum stress are horizontal and the mean stress is vertical. In such cases, the displacement remains essentially parallel to the strike of the fault.

These are also known as transcurrrent,, transform, wrench or tear-faults. In case of tear faults the strike of the fault is transverse to the strike of the country rock but the displacement is along the strike of the fault plane. But, in case the strike of the fault plane is parallel to the strike of the adjacent rocks and the displacement is along the strike of the fault plane, it is known as a rift fault.

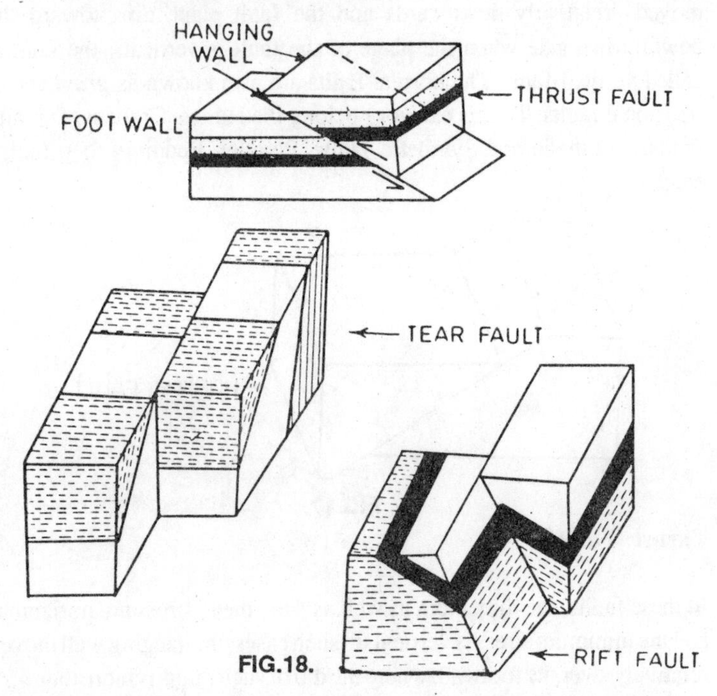

FIG.18.

Joint

Joints are fractures along which no significant displacement has occurred. They are found in most of the consolidated rocks of igneous, sedimentary and metamorphic origin. The tectonic mode of jointing is mostly due to large-scale twisting or torsion of the earth's crust involving both compressional and tensional forces. Joints divide the rocks into parts or blocks, with out any movement of the blocks past each other.

A rock may be traversed by a number of joints, but some of them may appear well developed and continuous for considerable length than the others. Such conspicuous joints are called *Master Joints* or Major *Joints*. A series of parallel joints is called a *Joint-set*. Two or more joint sets intersecting each other produce a *Joint-system*. Two sets of joints, nearly at right angles to one another, are said to form a *conjugate-joint system.*

Types of Joints

Genetically the following types of joints have been recognised in various types of rocks

(a) *Tensional joints* These are also known as *shrinkage-joints.* The polygonal cracks found on the dried-out mud flats due to contraction are the examples of Tensional Joints in sedimentary rocks.

In igneous rocks, such joints are developed due to cooling and contraction of the magma -mass i.e. the originally hot rock. The joints may be of radiating, irregular, hexagonal or polygonal type. Fine grained uniform rocks like basalt sometime show a remarkable kind of *Columnar-jointing.* The most perfect forms are composed of hexagonal columns which extend inwards from a cooling surface of the molten mass.

In granites and granodiorites several sets of joints may be observed, but commonly three set are prominent-one horizontal and two vertical at right angles to each other and to the horizontal set. When these sets are more or less equally spaced, the fracture planes give rise to cubical or rectangular blocks. Such a jointing is called *mural jointing.*

Tensional joints may also be due to deformation. Such joints are commonly developed along the axis of the anticlines, where the stretching effect is more pronounced.

Sometimes joints, more or less parallel to the surface of the ground, are developed in plutonic igneous rocks like granites as they are exposed at the surface. They originate mainly due to unloading effect, when the cover is removed through the processes of erosion. These joints are called *Sheet joints* or *exfoliation joints.*

(b) *Shear joints* These are also known as tectonic joints and are formed in a rock under compression. They form as a direct result of folding or thrusting in rocks. Two sets of shear fractures generally develop under compression and it has been established, through a number of studies that there exists much relation between the orientation of the joint patterns and the orientation of the stress system which operates in mountain building or continental uplift or local twists etc.

UNCONFORMITIES

An unconformity is a plane of discontinuity that separates two rocks, which differ notably in age. The younger of these rocks are nearly always of sedimentary origin and must have been deposited on the surface of the older rock, which is a surface of erosion. Therefore, an unconformity is regarded as a planar structure.

The formation of an unconformity may be attributed to three main processes like erosion, deposition and tectonic activity.Its development involves the following stages

(i) The formation of older rocks.

(ii) Upliftment and subaerial erosion of the older rocks.

(iii) The formation of a younger succession of beds above the surface of erosion.

An unconformity usually shows the following characteristics

(a) The overlying strata differ from the underlying ones with respect to their lithological composition, thickness and order of super-position.

(b) There is a difference in age, indicated by the fossil assemblages, between the overlying and underlying beds.

(c) The attitude (i.e. dip & strike) of the beds above the plane of discontinuity differs from those below it.;

(d) In most cases, a conglomerate horizon is present at the bottom of the younger set of beds.

Types of Unconformities

Four main types of unconformities have been recognised as follows:

(a) *Angular unconformity* In this case, the beds beneath the surface of erosion are folded or tilted so that there is angular discordance between the younger and older beds. Thus, the attitude of the rocks above and below the plane of discontinuity differs from each other. The contact is known as angular unconformity. Here, both the underlying and overlying rocks are of sedimentary origin.

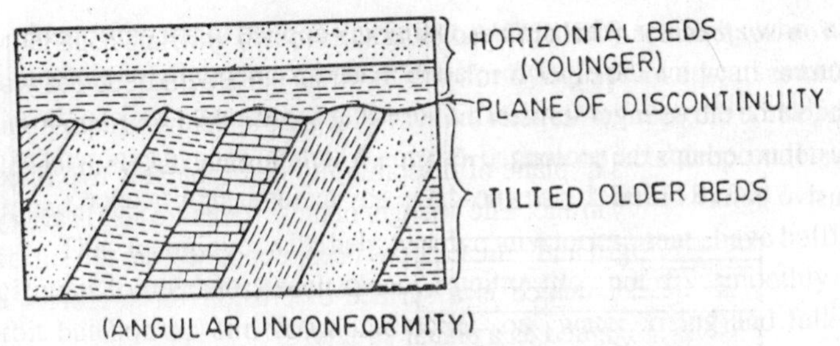

FIG.19.

(b) *Disconformity* It is also known as *Parallel unconformity* as the bedding above the below the plane of discocntinuity are parallel to each other. The lower as well as the upper series of beds dip at the same amount and in the same direction. This unconformity is developed when there is a lesser magnitude of diastrophism or disturbance between the deposition of the two succession of strata.

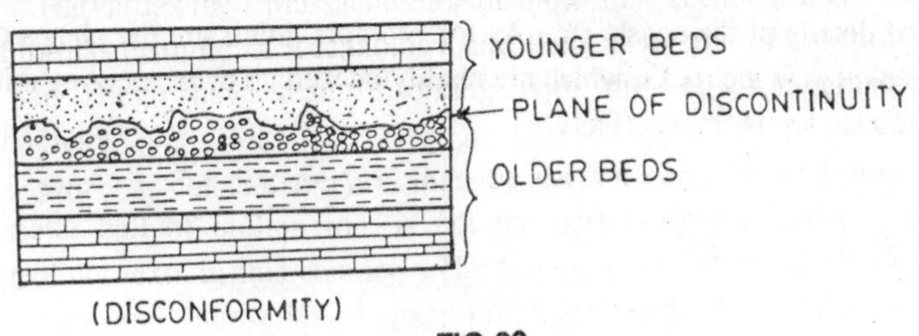

FIG.20.

(c) *Local-Unconformity* This is known as *non-depositional unconformity* It is similar to disconformity, but is local in extent and therefore the name. The time involved is also short, so the age difference between the overlying and the underlying beds is very less. Such an unconformity is also known as *Diastem.*

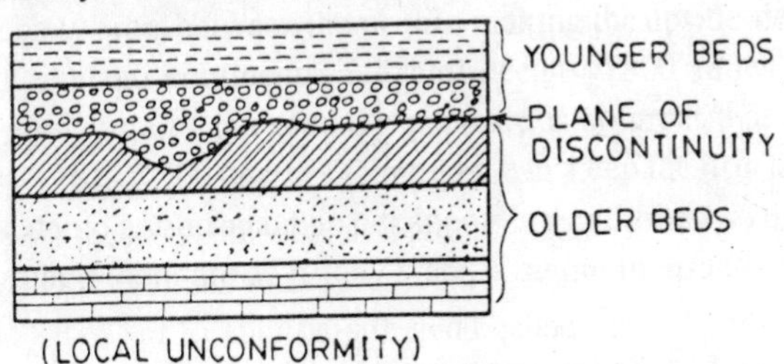

FIG.21.

(d) *Non-conformity* This term is commonly applied to unconformable structures in which the older formation made up essentially of plutonic rocks, while the younger formations are either sedimentary rocks or lava flows. It is perhaps the prolonged erosion which must have exposed the intrusive before burial. It does not have any tectonic significance.

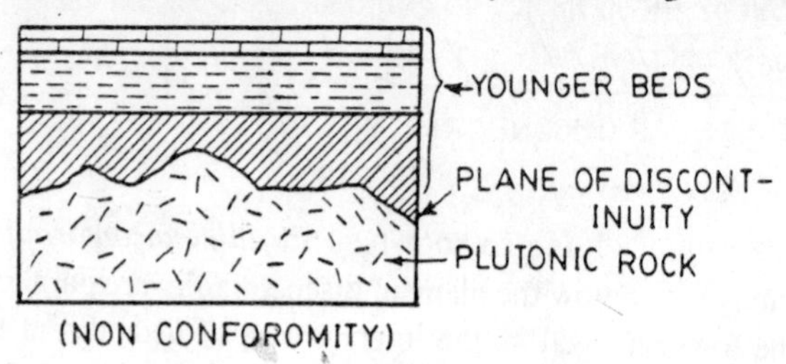

FIG.22.

The structures formed through tectonic movements often control the local pattern of drainage, the course of rivers, shaping the outlines and details of the coasts etc. As for example, joints are the planes of weakness in the rocks, which are readily opened up by weathering thus produce topographic effects.

10

DIASTROPHIC PROCESSES

The earth's crust is unstable, which is evident from the occurrence of earthquakes as well as of volcanic outbursts. Such events, eventhough, are seldom too frequent, they may produce new land from beneath the sea. These activities are related to the internal earth movements. Earth movements are of all kinds and degrees, ranging from those that cause a fracture involving a displacement measurable in centimetres to those that give rise to mountain ranges, raised continents and depressed seas, and oceans. These earth movements are as a whole, known as "diastrophism".

The term *'diastrophism'* has been defined as those crustal movements, gentle or severe, continuous or periodic, which cause the rocks to shift vertically, horizontally or in any other directions.

The diastrophic processes are broadly classified into two types as (1) Epeirogenic movement and (2) Orogenic movements. While the epeirogenic movements include the earth movements, which are predominantly vertical, causing elevation or depression of extensive regions of the earth's crust without much deformation; orogenic movements include the horizontal earth movements acting more or less tangentially to the surface by which the rock are crumpled and folded along narrow-belts resulting in the formation of great mountain ranges such as the Himalayas, Alps and Andes. It is, therefore, also known as 'mountain-building' movement.

EPEIROGENY

This type of tectonic movement is also termed as *'Continent-building'* movement as well as *'Oscillatory-movements'*, because of the probable change of the sign of movements when the elevation or uplifting is replaced by subsidence and the other way round. They occur even in the present times and had been taking place in the past geological periods. Evidences of elevation and subsidence are best displayed on and near the coastline.

GEOLOGICAL EVIDENCE OF ELEVATION

Eventhough, the sea-leval is not permanently fixed or indeed horizontal over the whole surface of the oceans, the sea itself often leaves indications of its former position. Sometimes sea-shells are found within the sand much inside the land-surface giving evidences that the actual beach of the past and the land has risen. Such uplifted beaches are found in many parts of Great Britain. Thus marine sediments found on land far away from the coast indicates elevation of land.

Uplifting of the earth's crust has been noticed in some parts of Northern Sweden and Finland that adjoin the Gulf of Bothnia. This is made obvious by the location of the port structures, a part of which is found to be at a remote distance from the sea.

An oft quoted example of elevation is furnished by the temple of Jupiter of Serapis, near Naples in Itlay. Here holes have been bored in columns about 6 metres above the floor of the temple by marine gastropods. It gives the evidence that there was a subsidence, after the Romans built the temple, beneath the sea when the boring animals bored the columns. Again there occurred the elevation which uplifted the columns above the sea-leval, where it stands today.

Reef-building corals cannot stand expoure to the sun and air for more than a few hours. Therefore, a coral reef found above the sea gives proof of change in its level.

The examples of elevation in India is mostly noticed in the Gujarat-coastal regions. The forms of the coast-line sometimes indicate the elevation of land.

The record of epeirogenic movement which had taken place in the past-geological time is conspicuously noticeable from the strata exposed in the Grand Canyon of the Colorado river in Arizona. During the millions of years represented by the strata exposed in the canyon the region was beneath or just at the sea level most of the time. This indicates that the earth's crust has been affected remarkably by epeirogenic movement in that region.

Geological Evidence of Subsidence

It is somewhat difficult to detect and prove that subsidence has really occurred as the evidences of the former position of the sea is destroyed or submerged beneath the sea water. The following are some of the evidences of subsidence:—

A considerable part of the country Holland is at present below the sea-level. Earlier this inhabited territory was above the sea level and now long, high dams have been constructed to protect it from the incursion of the sea. The rate of subsidence in the present times has been estimated to be 0.5 to 0.7 cm, approximately per year.

The remains of forests and peat beds which must have fluorished on land high above the present tide marks are sometimes found at or below the sea level. The submerged forests off the coast of Bombay and Pondichery, are the examples of subsidence of the coastal region of India. Peat beds in the Ganges delta can also be cited as examples of subsidence of a segment of the earth's crust.

The 'Temple of Serapis', in Italy near Naples had been showing evidences of subsidence, in recent years, at the rate of 2 cm approximately, per year.

Apart from the above, the following also gives evidences of recent subsidence. Coralreefs submerged far beneath the sea-level, estuaries, indented river valleys, lagoons as well as archeological findings substantiate a subsidence. For instance, Dwarka an ancient town in the Gujarat Coast of India, has been found submerged beneath the sea water, during archeological investigation.

Features of Epeirogenic Movements

(a) The most characteristic feature of the epeirogenic movement is that there is no crumpling of the rock-beds. The beds remain nearly horizontal.

(b) These earth movements affect areas of wide areal distribution.

(c) The periods of epeirogenic movements are quite large.

(d) These movements are reversible in nature. The same area may undergo upheaval followed by subsidence and vice versa.

(e) They affect the thickness of the sedimentary series, being formed, at the time of their operation.

CAUSES OF EPEIROGENIC MOVEMENT

The epeirogenic uplifts are thought to be caused due to a phenomenon known as *Isostasy*, which refers to the state of hydrostatic balance of the crustal segments of different thickness and density. As we know, due to weathering, erosion, transportation and deposition, the loads of rock on the surface of the earth are shifted from one place to the other. This disturbs the isostatic equilibrium. Usually the load is shifted on

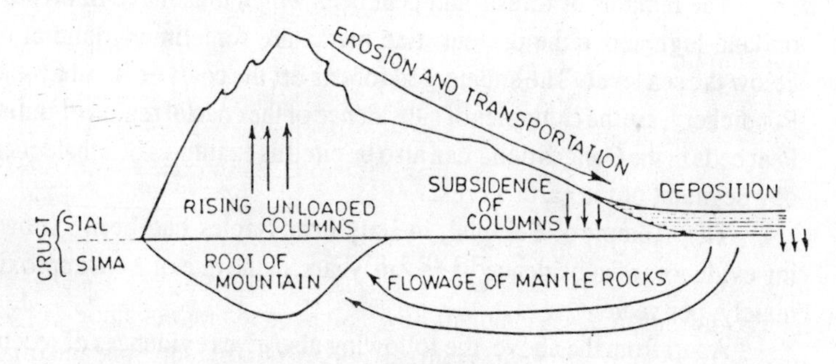

FIG.23.

the earth's surface from high relief features to lower areas. In order to restore the isostatic balance upset by the shifting of load, the mantle rocks (which always show plastic behaviour) move towards the area from which mass of rocks has been removed, at the bottom of the crust or beneath it, tending to elevate it. Isostatic movements are always vertical. The isostatic movements are also thought to be significant in areas which were earlier the sites of accumulation of large quantities of ice and snow, but later on there was melting away of ice as the glaciers retreated and the weight was suddenly reduced upsetting the isostatic balance. The state of isostasy is then re-established by compensating uplift of those areas.

The epeirogenic processes are rather slow but continuous and affects large portions of the crust. The sea-leval is mostly used as a datum-line to observe and measure the elevation or subsidence of the lands caused due to epeirogenic movements. Changes in position of the sea-level relative to the land are called 'eustatic changes', which are either due to the changes in the capacity of the sea-basin and or in the volume of the sea water.

OROGENIC MOVEMENTS

The word *Orogeny* has been derived from the Greek word *'Oros'* meaning mountain. The term was first introduced by the American Geologist G.K. Gilbert in 1890 to describe the processes of mountain-building. The orogenic movements are due to the horizontal forces in the earth's crust acting more or less tangentially to the surface, as a result of which rock-strata are crumpled and folded. This crumpling often gives rise to a narrow belt of folded strata, forming what is known as a mountain-chain. In contrast to the epeirogenic movements, orogenic movements are periodic in nature and severe in their effects.

Mountains, as is usually understood, are isolated or inter-linked masses of land elevated appreciably above the average altitude of their surroundings and are marked with pointed or ridge-like tops, called *peaks* .

The smallest unit of an elevated landmass is known as a hillock or mound, whereas the larger ones are called hills or mountains. The heights of mountains are commonly above 1000 metres while those of smaller dimensions are considered as hills. Mountains when maintain an echelon arrangement, they constitute a *range*. A number of ranges, formed more or less during the same time-period and existing approximately in a parallel fashion is called a *system*. A few systems and ranges, having a common trend irrespective of their time of formation, together form a *chain.* Several mountain chains combine to form a *Cordillera.*

Types of Mountains

Mountains are broadly classified as

(i) *Mountains of Accumulation* ; which are due to accumulation of volcanic materials during igneous activities or are due to accumulation of piles of sands forming sand dunes etc.

(ii) *Relict or residual mountains* ; which are mainly due to differential erosion in regions composed of of rocks of various strength and durability.

(iii) *Tectonic mountains* ; form by far the most important group of mountains. Folding, faulting and uplift play their roles together in mountain-building. Accordingly, tectonic mountains have been classified into three types as— (a) *Fold mountains* (b) *Fault mountains* and (c) *Dome mountains.*

While structures like *horst* and *graben* produced due to faulting and dome-shaped upheaval of the land is caused by igneous intrusions etc. belong to the latter categories of mountains, the fold mountains are formed due to intense lateral compression. The term *orogeny* is usually applied to the processes responsible for giving rise to fold-mountains.

Fold-Mountains

Geographically, it has been seen that the fold mountain chains are not in haphazard arrangement; instead they form rather distinct belts which extend for thousands of kilometers in length and hundreds of kilometres in width. Fold mountains usually show the following characteristic features

(i) Fold-mountain ranges tend to occur as long, narrow belts running more or less continuously for great distances.

(ii) The belts are irregular and are usually curved.

(iii) Fold mountains are marked by signs of intense lateral compression. The rock strata are folded and contain low-angle thrust faults.

(iv) They all have a great thickness of sedimentary rocks compared with those of the same age in adjacent regions. Among the various types of sedimentary rocks, shallow-water marine sediments are commonly found in these belts. Apart from the shallow-water sediments like graywackes, there also occurs deep-sea sediments like radiolarian oozes.

(v) There also occurs massive granite intrusions (batholiths) along the trend of the fold-mountain ranges.

(vi) In every fold-mountain belt, the folding is less intense near the margins and more intense towards the central part of the belt. Besides, the thickness of the individual strata tend to thicken toward the centre of the belt. It is, thus, apparent that the sedimantary strata exposed in mountain-belts were deposited in elongated trough-like depressions called geosynclines.

(vii) The youngest mountain-chains are situated along the continental borders.

Possible Causes of Orogeny

A number of theories have been suggested to explain the origin of orogenic movements. Some of the important theories are as described below:

1. *Theory of Contracting Earth* According to this theory, the earth in the beginning was a mass of hot and molten material and has attained the present position due to gradual cooling and solidification. The crust of the earth was conceived as a thick, strong, rigid layer resting on a molten or quasi-molten interior. As the interior of the earth cooled its volume decreased and it shrank away from the crust. The crust was thus forced to accomodate itself to a sphere of diminished circumference. This was accomplished by wrinkling or folding of the crust as the skin wrinkles on a drying apple.

Demerits

This theory fails to explain the following facts:

(i) The radioactive substances produce heat through their disintegration and that there is not much cooling beneath the crust as has been envisaged in this theory.

(ii) Fold-mountains should have been distributed uniformly all over the surface of the globle, but, in fact, they occur along a few belts and do not show any regular disposition.

2. *Geosynclinal Theory* This theory was based on the study of the Appalachian mountains, conducted by American geologists. According to this theory, an enormously thick pile of shallow-water sediments had been deposited in a long narrow, oceanic trough called *Geosyncline.*

The floor of the geosyncline subsided progressively in order to accomodate more influx of sediments. With continued subsidence of the floor of the geosyncline, the sediments would meet the internal heat and expand in volume. Expansion due to heating would buckle and crumple the strata into folds and at the same time, cause them to rise as mountains.

Demerits

(i) But the view that the deepening of the geosyncline was due to the weight of the overlying sediments is considered to be erroneous.

(ii) This process, even though explains the uplifting of the sedimentary layers to form mountains, does not account for the compressional forces.

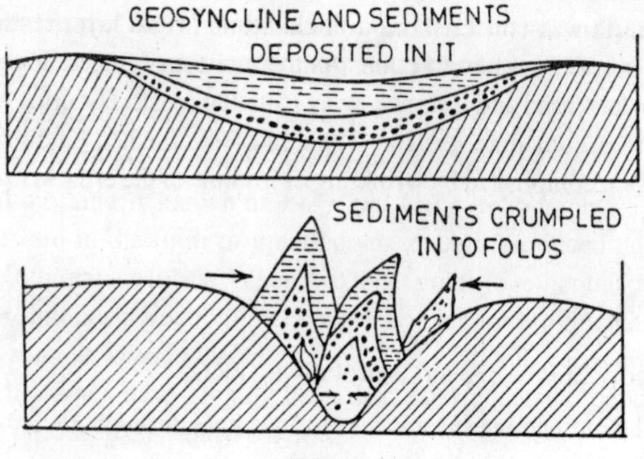

FIG.24.

Others believe that the geosynclinal basin on receiving the loads would sink down and in this process, the sides of the shallow trough would be brought nearer. This would generate compressional forces causing the formation of fold-mountains.

3. *Theory of Isostasy* Isostatic adjustment is believed to play an important role in mountain-building. The crust of the earth is considered to be in a state of gravitational equilibrium, (the heights of the various sectors vary inversely with the density of their component rocks). The process of adjustment between volume and density which tends to maintain the equilibrium figure of the earth, is known as isostatic-adjustment. This comes into operation when there is piling of great masses of rock materials or there is removal of large masses of rocks from any sector of the crust. It has been able to explain the increase in the heights of the Himalayas in the following way:

While active erosion lowers the Himalayas, the eroded material is deposited on the plain of the Ganges. Accordingly, the unloaded mountain tends to rise and the plain under the growing load of the deposits tend to sink.

But this process only accounts for vertical uplift and not the compressional forces. Besides, the process of isostatic adjustment is so sluggish that the isoslatic equilibrium is not perfectly attained.

4. *Continental-Drift Hypothesis* Continental-drift refers to the horizontal movement of the continents on a vast scale. According to the continental drift hypothesis,under the influence of the tidal force and the force generated by the movements of the earth's axis of rotation and revolution, the sialic layer of the crust was broken into pieces and began to move away from each other. The sialic blocks have been considered as light ships sailing on the ocean of Sima. It is also believed that the sialic blocks during their course of movement through the sima are affected by the resistance offered by the sima because of the different densities and conditions of moving blocks. Due to this resistance the frontal parts of the drifting blocks get crumpled and give rise to mountains. According to Wegener, mountains like the Himalayas, Rockies, Andes etc. have been formed in this way.

5. *Convection-Current Hypothesis* 'Convection' as we know, is a process of heat-transfer which particularly takes place in fluids and liquids. According to this hypothesis, it is belived that the mantle

(intervening layer between the crust and core) is made up of molten silicates and therefore in this medium heat is transmitted by convectional process. The convection cells are thought to be thousands of miles across.

This hypothesis suggests that the convection currents flow horizontally under the crust and exert a powerful drag on it. Two adjacent convection cells throw the crust into tension and pull the crust apart where they diverge and in to compression where they converge. Accordingly, orogenic belts are believed to form where the approaching currents turn down as shown in the following diagram:

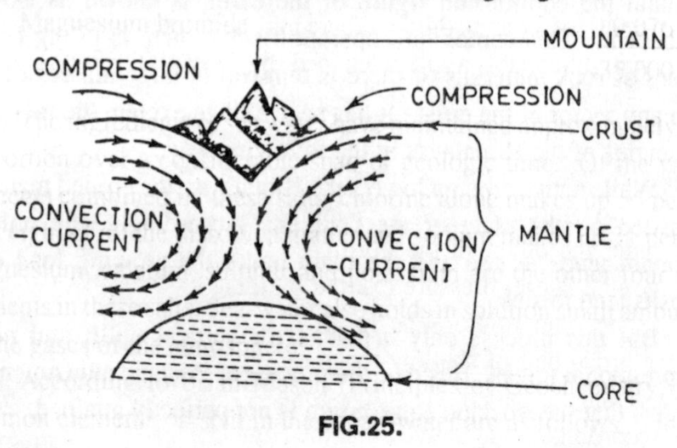

FIG.25.

Demerits

This hypothesis fails to explain why mountain-belt patterns are not symmetrical, while the convection cells in a relatively homogeneous earth are believed to be symmetrical. Besides, the materials deep beneath the earth are not quite suitable for the operation of convection currents in them.

6. *Concept of Gravity-tectonics* Some authors consider this concept as that of '*crustal sliding*', which takes into account the importance of the force of gravity in deformation of the crust. It has been assumed that in a geosynclinal depression, sedimentary beds occurring towards the axis are strictly horizontal in attitude whereas those formed along the fringes of the depression must have a gentle slope towards the axis. By virtue of their initial dip, due to the action of moisture and under the pull of gravity the marginal sedimentary layers gradually creep and slides down thereby producing folded structures (mountains).

7. *Concept of Plate-tectonics* It explains that the top-crust of the earth is a mosaic of several rigid segments called plates, which include not only the upper crust but also a part of the denser mantle below and are, therefore, considered as parts of the lithosphere. The lithospheric plates float on the plastic upper mantle known as the *asthenosphere*, a zone approximatly 150 kms thick that behave plastically because of increased temperature and pressure.

Plates may diverge, converge or move in parallels. Plates are said to diverge when two adjacent plates move apart and hot-magma comes up through the crack and solidifies, accounting for the formation of mid-oceanic ridges. But, plates are said to converge when two plates move toward each other and collide. There may be three types of convergence:in one case a plate capped by oceanic-crust can move toward another plate capped by oceanic crust; in the second case, an oceanic plate can converge with a plate capped by a continent (i.e. continental plate) and in the third case there may be a convergence between two continental plates. While in the first two cases one plate dives under the other, in the third case when two continental plates converge, the lateral pressure exerted by them crumples and compresses them into folds. The result is a mountain range and there occurs a broad-belt of shallow-focus earthquakes in that region. The mountain range is called the *suture-zone*.

Accordingly, it is belived that the Himalayas were formed due to the collision of the Indian-plate with the plate carrying the rest of Asia, due to the northward drifting of the former.

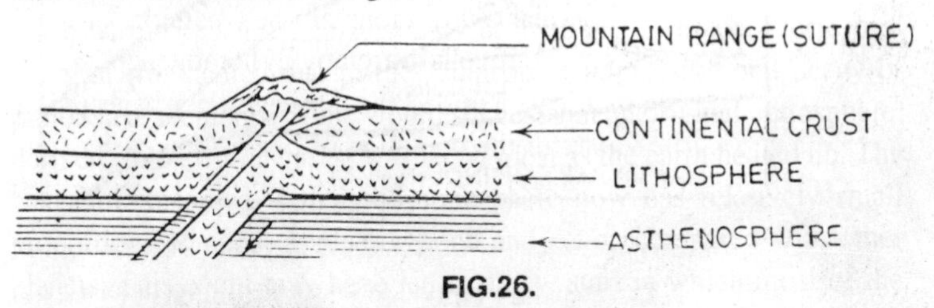

FIG.26.

All the above theories explain some of the possible modes of origin of mountain ranges and the causes of orogeny. The basić causes of orogeny still remains a controversial issue in geology.

Four major period of orogeny have been recognised as indicated below:

Name of Orogeny	*Geological Period*
Alpine or Himalayan Orogeny	Cenozoic
Hercynian Orogeny	Upper Palaeozoic
Caledonian Orogeny	Cambrian—Silurian
Baikalian Orogeny	Low Cambrian

Orogenic movements are also known as tectonic revolutions. From the study of the nature and characteristic features of fold-mountains, there, generally, appears to be three stages in an orogenic cycle as:

(i) *Initial stage* — Formation of geosyncline, subsidence and accumulation of sediments.

(ii) *Second stage* — Sever crustal movements causing folds, faults in sediments and igneous intrusion.

(iii) *Final stage* — Upheaval of the deformed sediments to form mountains.

Orogeny causes more crustal deformation in comparison to epeirogeny , even though both the processes together account for all diastrophic activities of the earth.

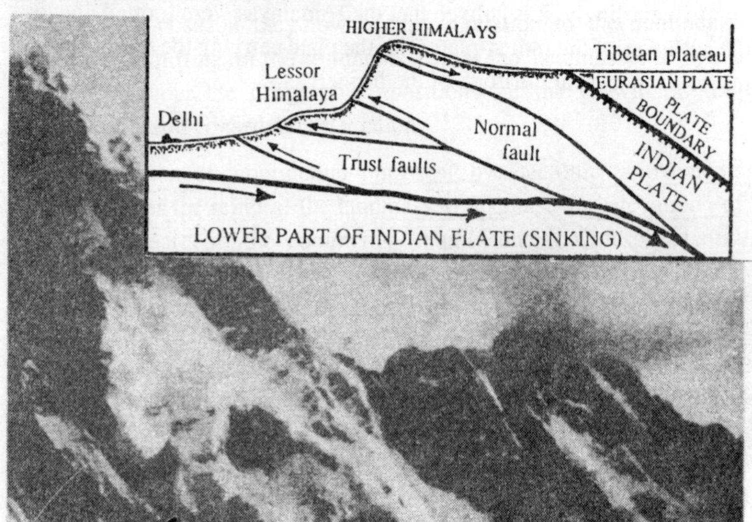

FIG 27 Himalayan Faults: cutting mountain to size.

The recent studies on the question "Are the Himalayas growing shorter?", has brought in to light a few new dimensions about the mechanism of mountain building.

Geologist know a lot about how mountains are built. But occasionally some new piece of evidence turns up that requires them to think again. The latest awkward fact concerns a series of faults high in the Himalayan mountains which do just the opposite of what geologists expected. They also shed some light on a possibly perverse question: why isn't Mount Everest even taller than it is?

The story of the Himalayas begins about 45 million years ago when the Indian and Eurasian continents first collided. Instead of both continents buckling together like an accordion, the Indian continental plate, a great slab of the earth's crust and mantle, started sliding under the Eurasian plate. Since then the Indian plate has been plunging under-neath the Eurasian one, but not without protest. The lower part, which is more dense, has split off from the top so that it can continue to sink. Meanwhile, the upper part has been breaking up. A series of roughly parallel fractures called "thrust faults" have divided it into a series of wedges. Because the two continents are still converging, the wedges have gradually been pushed upwards and southwards, piling up on one another and making the earth's crust in the region shorter and thicker. Hence the Himalays.

Recently, researchers working high in the Himalays found a series of "normal faults" which seem to spoil that story. Unlike thrust faults, which move the lower parts of the earth's crust upwards, normal faults move younger, higher material downwards. Thrust faults are found where the crust is shortening, making it thicker; normal faults occur where it is stretching, making it longer and thinner.

When geologists see a thrust fault and a normal fault in the same mountain range, they usually take it that they date from different periods. But detailed studies in the Himalayas have convinced Dr. Clark Burchfiel, Dr. Leigh Royden, and Dr. Kip Hodges of the Massachusetts Institute of Technology (MIT) that the Himalayan thrust faults and normal faults were active simultaneously.

They think this is because the Himalayas were getting too tall. They liken the earth's crust floating on the mantle to a rubber duck floating on water. If the duck were to become taller, it would float higher in the water until finally it became unstable and wobbled. When the earth's crust gets thick enough, it too becomes unstable. But mountains cannot easily tip over, because they are, as mountainers will have noticed, wider at the bottom than they are at the top. Instead they limit their height. Generally, mountains are worn away by erosion; the taller the mountain, the more rapid the erosion. But the rate of erosion of the higher Himalayas is probably not fast enough to keep the height of the mountains under control. The Indian and Eurasian continental plates are converging and thrust faults are piling up too fast.

According to the three geologists at MIT, the normal faults may be the result of a growth rate that erosion cannot handle, Because of the normal fault, a wedge part—way up the moutain range is being squeezed out towards the south, allowing the top to become lower as the wedge moves out from beneath the range. This is happening just where the researchers' mathematical studies of stress say it should- along the mountain's steepest slope. Thus although the Himalayas are still being built up from the bottom, they get no taller because they are collapsing at the top.

This adds a new twist to traditional theories of mountain-building. Geologists are now scattering over the globe to look for thrust faults and normal faults that happened at the same time; and they are rethinking the faults they observed in the past. On closer examination, it may turn out that mountains elsewhere have, like the Himalayas, reached their peaks.

11

VOLCANISM

The word *volcano* has been derived from the island of *Vulcano*, which lies off the north-east coast of Sicily. A volcano is essentially a conical or domeshaped hill or mountain, formed by the extrusion of lava or any other pyroclastic materials from a vent. Volcano is considered to be the most significant landform created by volcanism. *Volcanism* is the general term used to cover all the phenomena related to the eruption of magma to the surface of the earth. It is one of the most important evidences of the dynamic nature of the earth and arises from the forces which are endogenous in nature.

CAUSES AND FORMATION OF VOLCANOES

As it has already been indicated, a volcano is the result of the process of volcanism in whic' lava is extruded on the surface of the earth. This process is also known as *effusive magmatism*. Volcanism is considered to be the outcome of the release of high pressures which build up within the magma-chambers below the ground surface. The origin of volcanism can be dealt with the explanation for:—

(a) the origin of magma with its high temperature;

(b) the origin of volcanic gases; and

(c) the extrusion of magmas.

Magma may be produced because of the factors like—

(i) Geothermal gradient i.e. the increase of temperature with depth;

(ii) Accumulation of radioactively generated heat;

(iii) Relaxation of pressure locally etc. As we know, because of the high pressure at depth, the subcrustal region is in a viscous state. Any release of pressure due to some diastrophic movements melt the rock below and thereby producing pockets of magma.

It is also belived that water percolating through the crust gets converted in to steam due to increase of temperature with depth. Besides, a number of gases may be produced due to the effect of magma on the surrounding rocks. The steam and other gases may force the magma upwards causing eruption. There are instances when the eruptions are accompanied by powerful explosion of gases.

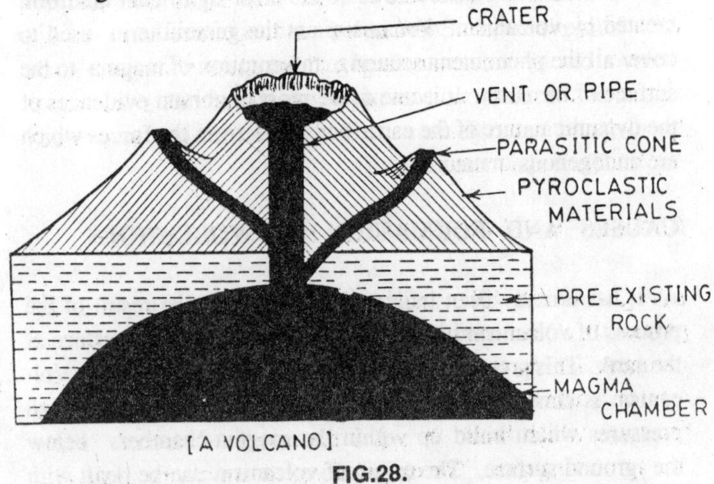

FIG.28.

The magma once produced will find its way upwards by the pressure of the overlying rock through fissures, joints, cracks etc, the development of which is mostly associated with crustal deformation. Reaching the surface, igneous material may pour out in tongue-like lava-flows or may be ejected as tephra (pyroclastic debris) under pressure of confined gases. Whenever, the magmatic material is ejected from an opening it spreads around the outlet and gradually cools and consolidates. The process of repeated eruption, cooling and consolidation over a period of time gives rise to a conical structure, commonly known as a volcano.

Every well-developed volcanic cone has near its top, a funnel-shaped depression that acts as the avenue for the magmatic materials to rise. This depression at the top of a volcanic-cone is known as *crater*. The crater is connected with the magma reservoir at the bottom through a pipe-like conduit, called volcanic pipe. Sometimes lava may also issue from the sides of the volcano, giving rise to secondary, satellitic or parasitic cones on the flanks of the main structure.

VOLCANIC PRODUCTS

Volcanoes usually produce three types of materials viz. solid, liquid and gaseous.

(a) *Solid products* Enormous quantities of solid materials are thrown out by volcanoes during an eruption. They consist of fragments of rocks or pieces of already cooled lava. The ejection of the solid materials are usually accompanied by violent explosions. The solid materials, during the initial stages of volcanism, mostly contain the fragments of the crustal rocks through which the pipe of the volcano passes; but at later stages they consist mostly the fragments of solidified lava, resulted from the partial solidification in the molten reservoir beneath the surface as well as the solidified lava of earlier eruptions. The rock fragments ejected during volcanic-eruptions are called *pyroclasts or tephra.* Generally, larger fragments fall at the edge of the crater and slide down its inner and outer slopes, while smaller ones are thrown into the surrounding plains or pile up at the foot of the cone.

According to their size and shape the pyroclastic materials

are classified as follows:

(i) *Volcanic blocks* These are the largest masses of rock blown out. These are either the masses of the solidified lava of earlier eruptions or those of the pre-existing rocks. They are usually angular and the diameter of the fragments is always above 32 milimeter. Thus they are the huge solid fragments ejected during a volcanic activity.

(ii) *Volcanic bombs* These are rounded or spindle-shaped masses of hardened lava, which may develop when clots of lava are blown into the air and get solidified before reaching the ground. Their ends are twisted, indicating rapid rotation in the air while the material was plastic. Because of their somewhat rounded appearance, they are known as volcanic bombs. The diameter of these fragments are always above 32 milimeter.

Bread-crust bombs are those volcanic bombs which present a cracked surface, may be due to the approximately solid state of the material from which they have been formed, which gives the appearance of the crust of a bread.

(iii) *Cinders or lapilli* The size of the fragments is between 4 mm to 32 mm, and are shaped very much like bombs. The term 'lapilli' is used when the fragments are not conspicuously vesicular; and in case of vesicular frag ments they are known as cinders. Still smaller fragments are called volcanic-sand.

(iv) *Ash* These particles range in size from 0.25 mm to 4mm and as such, are the fine particles of lava.

(v) *Fine-ash or volcanic dust* These are the minute pyroclastic materials, and their diameter is always less then 0.25 mm. In many instances volcanic dust was carried by wind to enormous distances and scattered over a vast territory forming volcanic dust layers.

Pyroclastic materials accumulating on the slopes and adjoining areas of a volcano with gradual compaction and cementation gives rise to rocks called *Volcanic-tuffs*. These tuffs when consist of angular fragmental materials, they are known as *Volcanic-breccia*; when volcanic bombs are predominant in the tuffs, they are referred to as *Volcanic-agglomerates*. The diameter of these fragments are always larger than 20 mm. The welded tuffs are commonly known as *Ignimbrites*. In certain instances, a great cloud of superheated vapours and incandescent rock material and volcanic ash are violently emitted during the eruption. These are called *Nuées ardentes* and are sometimes referred to as *glowing avalanches*.

(b) *Liquid products* Lavas are the major and the most important liquid product of a volcano. As we know, the magma that has flowed out on to the surface is called lava. All lavas contain gases, but because of the high pressure that prevails in the interior of the earth the content of gases and vapours in the magma is more. According to the composition and the gas content, the temperature of lavas during eruptions usually ranges between 900°C to 1200°C. Like magma, lava is also divided in to three types viz. acidic, medium and basic, depending on the silica content.

Acid lavas contain a high proportion of silica, have a high melting point and are usually very viscous and therefore their mobility is low. They cool very slowly and contain many gases in a dissolved state. They congeal at relatively short distances from the crater. Rhyolites, composed of orthoclase feldspar and quartz are the examples of acid lavas.

The lavas of intermediate or medium composition have the silica content between 55 to 60%. Andesite lavas are the best examples of the lavas of intermediate nature and they mostly characterise extrusions around the margins of the Pacific.

The basic lavas contain low percentage of silica, which is usually 50% or less. These lavas melt at lower temperature, and have a high density as well as liquid consistency. They cool quickly and contain little gas. These lavas are highly mobile and spread over large distances, forming flows or sheets. Basalts are the best examples of the basic lava.

Since the lava behave differently depending on their chemical composition they give rise to different configurations when consolidated, as described below:

(i) *Lava tunnels* Sometimes the outer surface of the lava flows; cools and solidifies first forming a crust while the lava is still in a liquid state inside. This enclosed liquid may drain out through some weak spots of the solidified flow forming a tunnel called a lava-tunnl.

(ii) *Block lava* It is also known as *aa*—lava. In this case, the gases escape explosively from the partly crystallised flows thus break the congealing crust in to an assemblage of rough and uneven blocks. The escape of gases increase the viscosity of the lava and helps in rapid cooling, giving rise to a solidified lava flow with spiny, rubbly surface. It is therefore the Hawaiian name, aa (pronounced ah-ah meaning rough or spiny) is applied to this type of lavas.

(iii) *Ropy-lava* Lavas with low-viscosity remain mobile for a longer period. These lavas usually contain much entrapped gas and cool very slowly. The lava spreads out in thin sheets and congeals with a smooth surface which wrinkles or twisted into ropy form like that of a stream of flowing pitch. It is also called Pahoehoe-structure.

(iv) *Pillow lava* Lava erupted under water-logged sediments in sea-water, beneath ice-sheets, or in to rain soaked air, characteristically emerges as a pile of rounded bulbous blobs or pillows. Basic lava of spilitic type often presents pillow structure.

(v) *Vesicular or Scoriaceous structure* When lavas heavily charged with gases and other volatiles are erupted on the surface, the gaseous constituents escape from the lava, due to the decrease of pressure, giving rise to a large number of empty cavities of variable dimensions on the surface of the lava-flows. Due to the presence of vesicles or cavities, the resulting structure is known as vesicular-structure. These cavities when filled up subsequently with secondary minerals, the structure is called *amygdaloidal structure* and the infillings as *amygdales.*

A highly vesicular rock, which contains more gas space than rock, is known as 'Scoria'. In more viscous lavas, when the gases cannot escape easily and the lava quickly congeals, it forms Pumice or 'Rock -froth', which contains so much void space that it can float in water.

(vi) *Jointing* As a consequence of contraction due to cooling joints are developed in the lava flows, which may be manifested in the form of sheet, platy or columnar structures.

(c) *Gaseous Products* Volcanic activity is invariably associated with emanation of steam and various gases from the volcanoes. Water vapour constitutes about 60 to 90% of the total content of the volcanic gases. Second in abundance to steam among volcanic gases is carbon-di-oxide. Amongst other gases which have been detected in considerable quantities, hydrochloric acid, sulphuretted hydrogen, sulphur-dioxide, hydrogen, nitrogen, boric-acid vapours, phosphorous, arsenic vapour, argon, hydrofluoric acid etc. are the most important.

The vents emitting sulphurous vapours are called '*Solfataras*' when carbon-dioxides are emitted they are called 'Mofettes' and in the case of emission of boric-acid vapours, they are known as Saffioni.

TYPES OF VOLCANOES

Volcanoes are classified in to various types on the following bases

(a) Continuity of eruption;

(b) Nature of eruption; and

(c) Mode of eruption;

(a) *Continuity of eruption* On this basis volcanoes are classified as—Active, Dormant and Extinct. The active volcanoes are those which still erupt either; intermittently or continuously. The volcano which has not erupted for a longtime but are expected to be active at any time is called a dormant volcano; whereas an extinct volcano is one which has stopped eruption over a long time. Killimanjaro in Tanzania (Africa) is considered to be a dormant volcano.

(b) *Nature of eruption* Depending on factors like chemical composition of lavas, the amount of gas contained in them and their pressure, temperature etc. the volcanic eruptions may be quiet, intermediate or violent.

In quite-type of volcanoes, the lava erupts quietly without any explosion. In this case the lava is of basaltic composition, which is highly fluid and holds little gas.

Intermediate-type of volcanoes erupt intermittently with explosion in the beginning and gradually the explosive action dies down and lava is emitted quietly.

In the Violent-type of volcanoes, there are explosive eruptions. The lava in such cases are of acidic nature and have high degree of viscosity. These explosive volcanoes usually produce huge quantity of pyroclastic materials.

(c) *Mode of eruption* On the basis of the mode of eruption volcanoes are classified as—Central-type and Fissure-type.

1. *Central type* This type of Volcanoes are represented by a cone crowned by a bowl-like depression called the crater and a vent, connecting the crater with the magma-chamber, through which the eruption products reach the surface. A number of central types of volcanoes have been recognised depending on the chemical composition of the lava, gaseous contents and the nature of the volcanic structure, as follows:-

(i) *Hawaiian type* In such cases the lava begins to pour over slowly the edge of the crater and flow down the slope. Thus there is silent effusion of lava without any explosive activity. Sometimes, the lava, foamed by gases, is sprayed into the air and solidified in the form of long glassy threads known as *Pele's Hair*. The lava is basaltic in nature. The Hawaiian type of volcanic eruption is characteristic of Mauna Loa and Kilauea on the Hawaiian Islands.

(ii) *Strombolian-type* In this type, the eruptions are rhythmic and they occur at intervals of 10 to 15 minutes. The lavas emitted are of basaltic composition which are less mobile and with more viscosity in comparison to those of Hawaiian types, because of more accumulation of gases. Moderate explosions occur with the eruption, ejecting volcanic bombs, lapilli and slags.

The Volcano 'Stromboli' in the Mediterranean Sea shows this type of eruption. Since glow from the ejected masses are visible afar to men on ships with regularity, it is called the *Light House of the Mediterranean*.

(iii) *Vulcanian-type* In this case, eruption takes place at longer intervals and the lava is more viscous which quickly solidifies between consecutive eruptions, producing explosions. Each new-explosion causes the shattering of the congealed cover. They emit much ash. This type is named after 'Vulcano' in the Lipari Islands, north of Sicily, which shows this type of eruption.

(iv) *Vesuvian-type* These are characterised by extremely violent eruptions of lava which is highly charged with gases, possessing a relatively high degree of viscosity, during a long period of superficial quiescence. The eruptions occur after long intervals, usually measured in tens of years, ejecting huge amount of volcanic products in the form of volcanic ashes, lapilli, bombs etc. The lava flowing out of the crater runs down the slope of the cone. After the volcanic activity subsides, it remains at rest for an indefinite time.

Volcanoes like Vesuvius, Etna located in the Mediterranean shows this type of eruption; hence of the name.

(v) *Plinian-type* These are the most violent type of Vesuvian-eruption. In such cases, huge quantities of fragmental products are ejected with little or no discharge of lava.

(vi) *Pelean-type* This is the most violent type of all the eruptions. Here the lava is of andesitic composition, and highly saturated with gases and possesses a high degree of viscosity. Such lava congeals in the crater and because the vent gets plugged, the free exit of gases is prevented thus creating a tremendous pressure beneath the plug.

Here the lava, therefore, force its way out through side fissures and sweeps down the slopes as avalanche of molten rock-materials of self-explosive type and gases. This combination of extremely hot, incandescent fine ash and coarse rock fragments permeated with hot gases is known as *Nuèès ardentes*.

The town of St. Pierre at the foot of Mont Pele'e was destroyed by nue'es ardentes, during the eruption of Mont Pele'e in 1902. This type of eruption is exhibited by Mont Pele'e located in the Island of Martinique in West Indies.

2. *Fissure type* Sometimes volcanic eruptions take place along a fissure or a group of parallel or closed fissures. Usually volcanic cones are not produced through fissure—eruptions. Lava, flowing out of fissures, spreads out over extensive areas forming lava sheets. Fissure - eruptions are characterised by quiet welling out of molten lava. The Deccan Traps in India are made up principally of basaltic lava-flows, which were erupted mostly through fissures and covered a major portion of the Deccan—plateau.

VOLCANIC TOPOGRAPHY

Volcanoes produce distinctive land-forms, through extrusion of lava. The most obvious landform created by volcanism is a volcano i.e. a conical hill composed of materials erupted during volcanism. The volcanic topography includes both positive as well as negetive relief features. The positive relief features consist of the hills, cones, mountains, lava plateaues etc. while the negetive relief features include the depressed forms like craters, calderas, volcanic-tectonic depressions etc.

Positive - Relief features Some of the important positive relief features produced by volcanism are as follows:

(a) *Hornitos* These are the terms applied to very small lava flows.

(b) *Driblet cones* These are small conelets produced by the most acid and viscous lava.

(c) *Cinder cones* These cones are formed due to the accumulation of loose rock fragments ejected during volcanism immediately round the vent. Since such volcanoes erupt only a small quantity of lava, the cones are built up principally of pyroclastic materials as cinder, bombs, ash etc. The sides of the cones have a slope that is the angle of repose of the pyroclastic debris. These cones commonly have slopes of 30° to 40° and do not exceed a height of 500 metres. As these cones grow around a central vent, with the accumulation of pyroclastic debris they

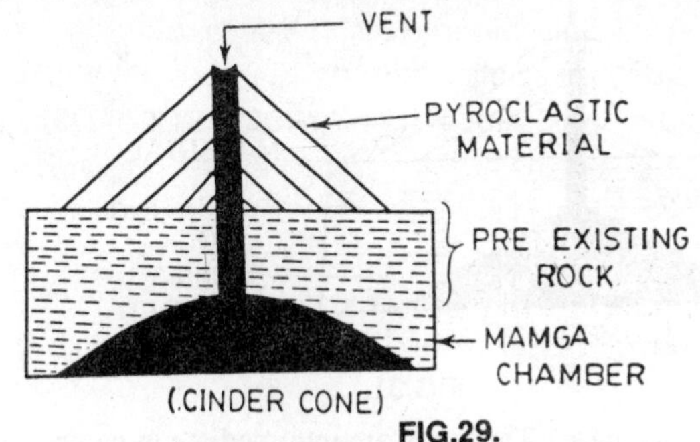

FIG.29.

are also known as *Ring crater* or as *tuff-ring* when composed dominantly of finer materials.

(d) *Lava cone* These are built up of lava flows, due to heaping of lava during quiet-type of eruption. Such cones are usually dome-shaped but not conical because of the fluidity of lava. The form of these domes varies depending on the nature of the lava, whether acidic or basic. They have gently sloping sides and very broad base.

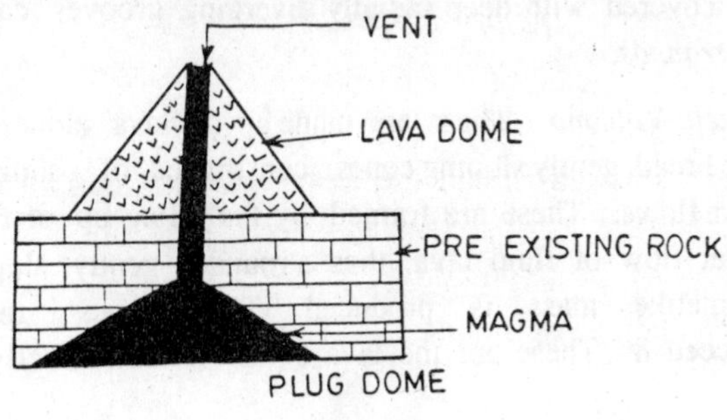

FIG.30.

These are also known as *lava-or plug dome*. Highly viscous silica-rich lavas form steep sided cones called cumulo-domes or tholloids.

(e) *Composite cone* The composite cones are made up alternately of pyroclastic material and lava flows. As indicated by its name, a composite volcanic cone is formed partly by explosive eruptions (which are chiefly responsible for increasing its height and steeping its summit slopes) and partly by lava that flows forth quietly and

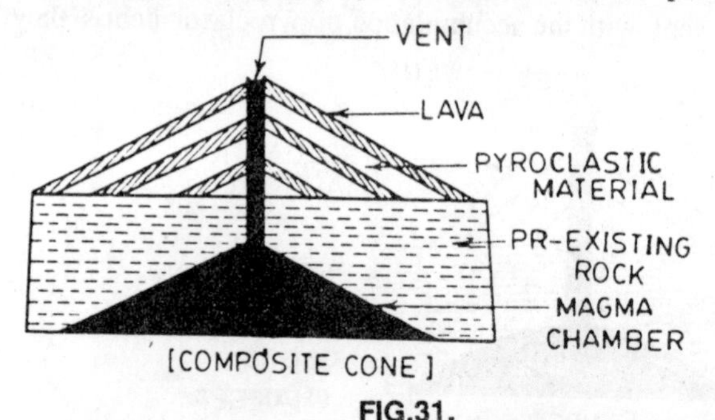

FIG.31.

widens its base. Thus they are intermediate in nature and composition between the cinder cones on the one hand and lava-cones on the other. These cones are built up over long spans of time. Due to the rude stratification of explosive materials and lava flows in such volcanoes they are also known as 'Strato-volcanoes'. The important examples of these cones are shown by Fujiyama in Japan, Mayon in the Phillipines, and Vesuvius in Itlay. The slopes of these volcanoes are sometimes found to be covered with deep radially diverging grooves called *barrancos*.

(f) *Shield-Volcano* These are made up of lava alone and are broad, gently sloping cones constructed of solidified lava flows. These are formed by the piling up of flow after flow of fluid lava; thus a rounded gently sloping domelike mass is produced. The slopes rarely exceed 8°. These are the lava cones formed chiefly of

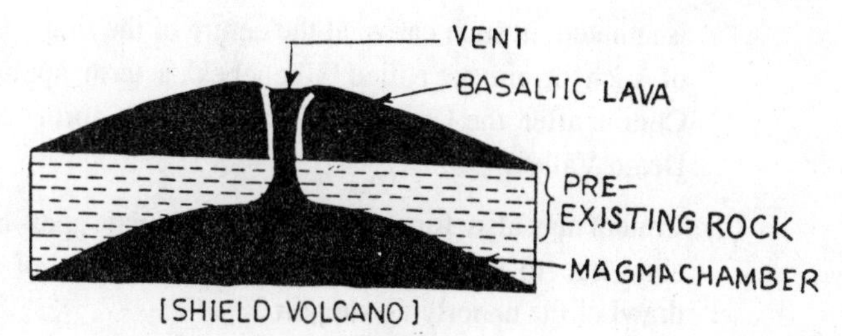

FIG.32.

basalt flows and are frequently the sites of fissure eruptions. The volcano *Mauna Loa* of Hawaiian islands is an excellent example of shield-volcano.

(g) *Spatter Cone* Sometimes small, steep-sided cones are developed on the surface of a solidified lava flow, at the spots where breaks occur in the cooled surface of the flow due to the pressure of the entrapped gases. Thus hot lava and gases blow out forming steep sided cones which rarely exceed 10 metres in height.

(h) *Volcanic -plateaus* These are built by extensive extrusions of lavas, notably basalts, and are associated with the fissure eruptions. There is piling of lava flows and each layer ranges in thickness between 15 to 100 metres and hundreds of square kilometres in extent. The Deccan plateau of the Peninsular India represents the best example.

Negetive Relief Features The negetive relief features in a volcanic terrain commonly consist of the following :—

(a) *Crater* This is a depression located at the summit of the volcanic cone. It is usually a funnnel-shaped hollow that marks the top of the volcanic-vent. These craters may be formed as follows:

(i) Sometimes pyroclastic materials ejected out during a volcanic eruption, accumulate around the opening i.e. at the upper end of the vent. Thus a ring-shaped mound is formed, which is known as the *Crater-ring* . The crater is situated, in such cases, at the centre of the ring. Craters of such origin are called 'Ubehebes', a term applied by Cotton after the Ubehebe Craters at the north end of Death Valley.

(ii) Craters may also form by subsidence of the apex of the cone due to collapse of the structure because of withdrawl of the underlying support.

(iii) In majority of the cases, craters are formed by the explosion that occurs during a volcanic eruption, whereby the apex of the volcano is blown out.

Sometimes due to explosive eruption small craters are formed without the building of cones. Such small craters are known as *explosion pits.* The explosion-pits, when occupied by small water-bodies form miniature crater-lakes. Such small crater lakes are seen in the Eifel district of Germany, where they are called *maare* or *maars*

(b) *Caldera* A caldera is generally a volcanic crater, which is enlarged to a diametre of several kilometres. These circular depressions commonly have steep inner walls and a flat floor. The calderas may have different origins and therefore, are of different types as follows:

(i) *Explosion Calderas* Sometimes due to violent volcanic explosion the entire central portion of the volcano is destroyed and a great central depression remains. This great central depression is called a Caldera. Bandai-San in Japan represents an example of this type of caldera.

(ii) *Callapse Caldera* These are formed when the top of a volcano collapses or subsides into a vacated magma chamber beneath the volcano, because of the withdrawl of the underlying support. It is believed that the rapid eruption of lava and pyroclastic materials etc. lower the level of magma in the main reservoir to such an extent that a potential void is left. Thus the crater-floor subsides, since there is no support from beneath. The caldera of the volcano Krakatau in Indonesia is an example.

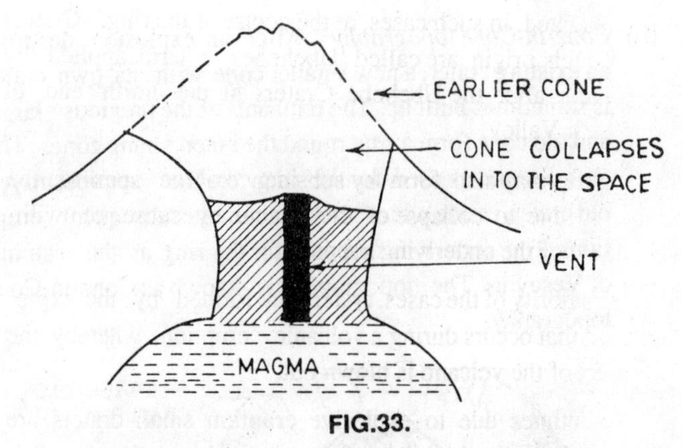

FIG.33.

(iii) *Resurgent Caldera* According to M.P. Billings, ('Structural Geology', Prentice Hall of India, New Delhi), these are formed where collapse is followed by the doming of the central block. Example—Valles Caldera in New Mexico.

(iv) *Erosion Caldera* These calderas are formed by the enlargment of craters of volcanoes by various eroding agents.

The Buldir-caldera is the largest known caldera located between the islands of Kiska and Buldir in the Aleutian chain.

(c) *Volcanic—tectonic depressions* Due to tectonic reasons sometimes depressions are formed on the slopes of the volcanic cones. These are volcanic-tectonic depressions.

IMPORTANT FEATURES ASSOCIATED WITH VOLCANOES

(i) *Volcanic-necks* It is produced when the conduit of the volcano is chocked up due to gradual consolidation of lava and because of their resistant nature they are preserved even after the former volcanic cone and domes are obliterated by erosion. The term *volcanic-plug* is used for the lava-mass solidified in the conduit of the volcano. These are also known as *Puys*.

(ii) *Cone-in-Cone topography* After an explosion destroys an existing crater, a new smaller cone with its own crater is sometimes built up. The remnants of the previous large and old cone form a ring round the inner young cone. The term *Somma* is used for the ring or the semidestroyed old crater, inside which rises a cone by subsequent eruptions. The term was first used for the ring at the summit of Vesuvius. The topography is known as Cone-in-Cone topography.

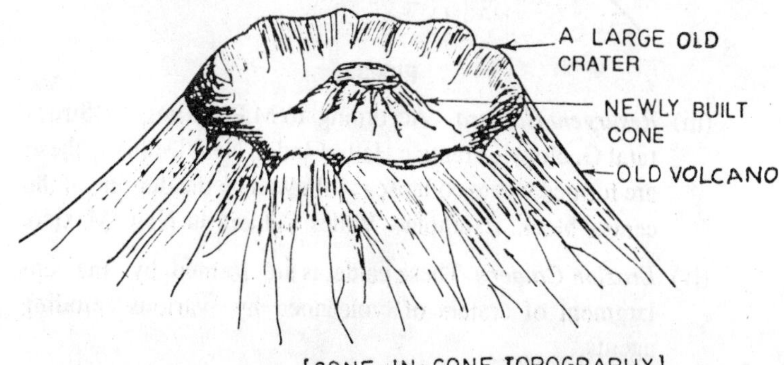

[CONE-IN-CONE TOPOGRAPHY]

FIG.34.

FEATURES ASSOCIATED WITH THE DECAYING PHASES OF VOLACANISM

Even after a volcano ceases to eject pyroclasic materials and lava, it may continue to emit steam, gases, mud etc. and these activities are expressed in form of fumaroles, hot springs, geysers, mud-volcanoes etc.

(a) *Fumaroles* These are fissures or vents through which volcanic gases are ejected. The emission of gases, especially steam, occurs in all stages of volcanic activity, but it becomes the dominating and characteristic features of its declining stage. A volcano discharging gases only is said to have attained the fumarolic stage. From the analysis of the materials accumulated by fumaroles, it has been established that the fumarolic gases belong to the groups of halides, sulphur, carbon, water-vapour, boric acid vapour, hydrogen etc. The gas composition of fumaroles largely depends on their temperature and accordingly they have been classified as follows:

(i) *Dry fumaroles* These are charaterised by high temperature (around 500°C) and no steam. They are highly saturated, with chlorous compounds of sodium, potassium and iron.

(ii) *Acid fumaroles* These are also high temperature fumaroles, where the temperature is between 300 to 400°C. But they contain steam, hydrogen chloride and sulphur dioxide.

(iii) *Alkaline fumaroles* These are within the temperature range of 200 to 300°C and contain mainly ammonium chloride.

(iv) Solfataras, are the fumaroles emitting sulpurous vapours and are characterised by the temperature of 100 to 200°C.

(v) Mofettes are the fumaroles which mainly emit carbon-dioxide and steam and are with a temperature below 100°C.

(vi) Saffioni, are the fumaroles emitting boric-acid vapours.

The gaseous emanations from the fumaroles by the reaction with the country rocks sometimes bring about remarkable change in their mineralogical composition. The processes as a whole, are known as *Pneumatolysis.* The unique fumarolic field is the "Valley of Ten Thousand Smokes" near Katmai Volcano (Alaska).

(b) *Hotsprings* These are also known as Thermas. These are avenues through which hot-water escapes to the surface. Hot-springs occur not only in volcanic regions but also in areas characterised by recent tectonic movements. Hotsprings bring chemical substances to the surface. Calcareous deposits formed from hot-spring are known as Travertine or Tufa. Siliceous deposits produced by hot-springs are called *Siliceous-sinters.*

The water of the hotsprings usually gets heated with the increased temperature below, may be due to magmatic or radioactive-heat.

(c) *Geysers* These are hot-springs ejecting boiling water and steam intermittently. In these cases, hot water and steam are explosively discharged. The waters of geysers contain a large amount of mineral matter, predominantly silica. At the time of eruption, the mineral matter fall out on the edges of the opening or vent, forming conical structures. Such mineral deposits are called *geyserites.*

(d) *Mud-volcanoes* These are cone-shaped mounds, built up of mud, like miniature volcanoes. It is believed that hot water when passes through mud or volcanic ash in its ascent to the surface, it becomes muddy and form a conical mound with a crater at the top. In certain cases the mud eruptions are quiet where the mud boils up and erupts in small spouts, whereas in other cases the eruption is explosive. The volcanic mud-flows are also known as *Lahars*. Mud-volcanoes occur in volcanic regions as well as in the areas of oil deposits(where the driving forces are gases produced due to decomposition of organic matter).

PSEUDO-VOLCANIC FEATURES

Certain topographic features of non-volcanic origin resemble volcanic forms and are therefore known as pseudovolcanic-features. They include the following:

(i) *Meteorite Craters* These are the depressions formed due to the impact of falling materials of large dimensions. The Lonar lack in Buldana district of Maharashtra was originally thought to be a crater lake, but was later confirmed to have been formed by the impact of a giant meteorite.

(ii) *Salt-plugs* As we know, under high pressure salt deforms plastically and behaves like an intrusive, deforming and piercing the overlying sediments. Salt extrusions may take the form of salt hills which exhibits many features of plug domes or lava cones with peaks and small sink-holes, which look like craters produced due to subsidence.

(ii) *Mud-volcanoes* Some of the mud-volcanoes are of non-volcanic origin. As for example, the volatile hydrocarbons given off from the petroleum-bearing beds beneath cause mud-eruptions, as in case of the mud-volcanoes at Baku on the Caspian, in southern Baluchistan, in Burma etc.

Apart from what has been described above, William D. Thornbury (Principles of Geomorphology, Wiley International Edition, Second Edition, New York 1985) has mentioned that craters formed by bombs and mine blast (Bomb and mine craters) have features of craters developed due to volcanic explosion.

DISTRIBUTION OF VOLCANOES

At present there are more than 500 active volcanoes but these are not scattered irregularly over the globe. It has been observed that volcanic activity is strictly confined to certain limited sectors of the crust. They are mostly found on the marginal parts of the continents and in the littoral zones of oceans and seas they are situated within the boundaries of young tectonically mobile mountain structures. Thus, the distribution of the present day volcanoes are restricted mostly to the mobile-zones of the earth's crust. Of course there are certain exceptions as for example, there are no volcanoes in the Himalayas and there is no sign of recent-folding in Iceland.

The most important belt of volcanoes is the so-called *Ring of Fire* or the *Cicum-Pacific belt* where more than 60 percent of the active volcanoes are located. It extends through the Andes of South America, Central America, Mexico, the Cascade Mountains of western U.S.A., the Aleutian Islands, Kamchatka, the Kurile Islands, Japan, Phillippines, New Guinea, Solomon Islands, New Hebrides and New Zealand. Volcanoes like Cotopaxi, Katmai, Fujiyama etc. are located in this belt.

Another belt runs north-south through the Atlantic and accordingly is known as Atlantic belt. It extends, from Jan Mayen Island in the north, through Iceland, the Azores and the Canary Islands to the Cape Verde Islands in the south. Volcanoes like Mt. Pelee, St. Helena, Mt. Hekla, Mt. Helgalfel etc. are located in this belt.

Another volcanic belt is the Mediterranean Himalaya belt extending east-west from the Alps via the Apennines to the Caucasus and the mountains of Asia Minor. Volcanoes like Vesuvius, Etna, Mount Ararat and those of the Lipari Islands are located in this belt.

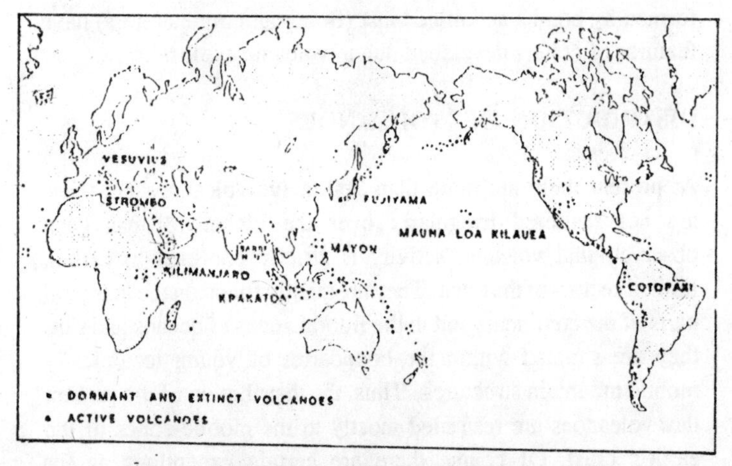

FIG.35. THE DISTRIBUTION OF VOLCANOES

Effects on Human beings Even though volcanoes cause a lot of destruction, some of its effects are also beneficial to mankind. In certain cases, it is possible to harness the underground heat generated by volcanism to human needs i.e. to meet the energy requirements of future. This is what is commonly referred to as *Geothermal Energy*.

Besides, volcanic materials weather quickly in warm and moist climates to provide fertile soil. These soils are rich in plant nutrients, as for example, the *Black-cotton soil* of Western India.

Chemicals like sulphur, boron etc. are mostly obtained from volcanic sources and volcanoes also provide good building materials.

EXAMPLES OF RECENT VOLCANISM

(i) Mount Chilcholan an active volcano started eruption on 29th March 1982. It is situated in the South-eastern region of Mexico.

(ii) Mount St. Helens, in southern Washington, started eruption in 1980.

(iii) Mt. Helgafell in the Westermann Islands, off the south-West coast of Iceland, erupted in 1973.

(iv) Mt. Etna, Sicily in Italy erupted in 1965.

(v) Klyvuchevskaya-Sredinny Khrebet, in the Kamchataka peninsula of U.S.S.R. erupted in 1964.

12

ISOSTASY

The word *Isostasy* has been derived from the Greek word *Isostasios* meaning equal standing or in equipoise. The theory of isostasy explains the tendency of the earth's crust to attain equilibrium and the distribution of the material in the earth's crust which conforms to the observed gravity values. This theory was developed from gravity surveys in the mountains of India, in 1850. The term was first proposed by Clarence Dutton, an American geologist in 1889.

This doctrine states that wherever equilibrium exists in the earth's surface, equal mass must underlie equal surface areas; in other words a great continental mass must be formed of lighter material than that supposed to constitute the ocean-floor. Thus, there exists a gravitational balance between crustal segments of different thickness. According to Dutton, the elevated masses are characterised by rocks of low density and the depressed basins by rocks of higher density. In order to compensate for its greater height these lighter continental material must extend downward to some distance under the continent and below the ocean-floor level in order that unit areas beneath the oceans and continents may remain in stable equilibrium. Accordingly, a level of uniform pressure is thought to exist where the pressure due to elevated masses and depressed areas would be equal. This is known as the 'Isopiestic-Level'.

Isostatic balance is upset if enough matter is transferred from one region to the other on the earth's surface. Intrusion of igneous material, accumulation of snow and ice, deposition of sediments etc. puts additional load whereas denudation, melting of ice etc. causes unloading. This process of loading and unloading disturbs the balance and the process of compensation takes place to restore it. With the removal of material from the top of the mountains through erosion, it becomes lighter. Materials should, therefore, move into the roots of the mountains at depth through the interior of the earth. This movement is termed as compensation which takes place in the form of elevation and depression. This is because, at depths, rocks apparently flow slowly outward from an overloaded area, that subsides to form a depression, to an underloaded area which gets higher and higher to form an elevation. The isopiestic level is the level of compensation.

The zone between the level of compensation and the surface of the earth is the zone of compensation or lithosphere. The zone below the isopiestic level is called the asthenosphere.

Three theories have been propounded to explain the concept of isostasy:

1. *Airy's Theory* He presumes that the crustal blocks are of equal density and unequal thickness. As such the blocks constituting the mountains are thicker than those on which the plains lie and therefore they stand higher up as is the case with the masses of ice floating in water. Floating ice is eight-ninths submerged and the higher the ice rises above the water level the deeper is the submerged portion. Thus the roots of the mountains sink in the basaltic substratum to depths proportional to the heights above.

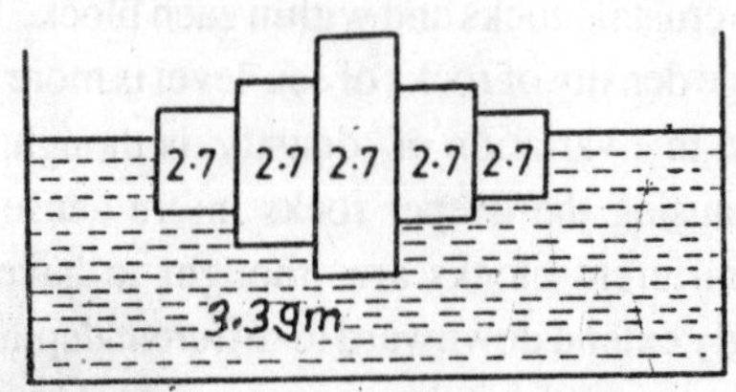

FIG.36. (AIRY'S THEORY OF ISOSTASY)

This has greater support from recent geophysical data. For example Mt. Everest in the Himalayas rises to a height of about 9 Kms. whereas right beneath it the crust is about 80 Kms thick.

Thus Airy suggested that blocks of the lithosphere had a constant density of 2.7 gm. per cubic centimetre and floated in the asthenosphere of density 3.3 gm per cubic centimetre.

2. *Pratt's Theory* According to this theory, there is a difference in the density of rocks in the crust and that the heights of the crustal blocks are determined by their densities. As such blocks made up of lighter material are at higher elevation than those consisting of denser material. Lighter material, has therefore, been assumed to lie under mountains and heavier material under ocean and there also exists a boundary, between the upper blocks and the lower dense rocks, at a uniform depth known as the level of compensation. Thus, the rocks constituting the elevated masses and depressed areas exert equal pressure at the level of compensation.

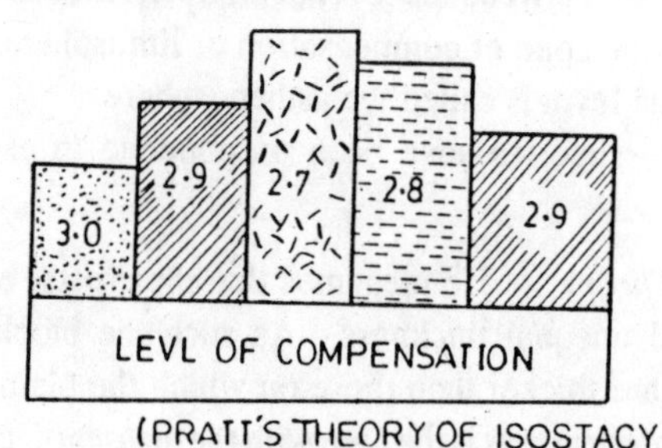

(PRATT'S THEORY OF ISOSTACY

FIG.37.

3. *Heiskanen's Theory* The assumptions of both Airy and Pratt have been combined in this theory. Here it is assumed that density varies both between crustal blocks and within each block. It has been observed that the average density of rocks of sea-level is more than those at higher elevations and this variation of density is thought to continue further downwards causing the deeper rocks more dense than the shallower ones. Thus different blocks are thought to have different densities and accordingly extend downward to different depths. It explains for the roots of mountains and for the variations in density in different parts of the crust.

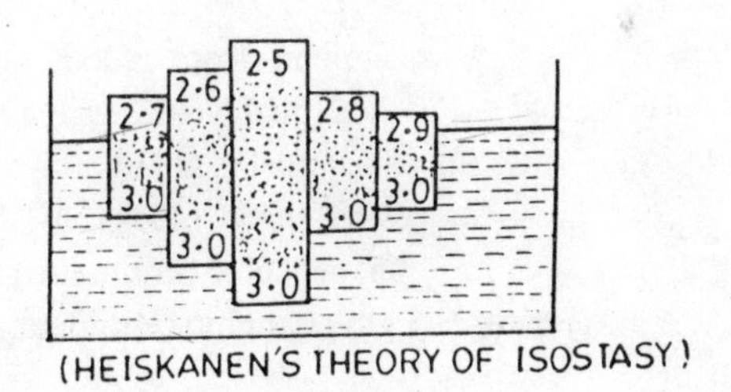

(HEISKANEN'S THEORY OF ISOSTASY)

FIG.38.

The theory of isostasy convincingly explains the vertical uplift of the mountains but it has not yet been possible to establish that isostasy is the factor initiating tectonic movements. The role of isostasy in the developments of the earth's crust is rather modest and not decisive.

The idea of isostasy is supported by the fact that the melting of ice from the glaciers in Scandinavia led to a reduction of load and the consequent rise of the area. The theory of isostasy is also confirmed by the seismic data.

13

CONTINENTAL—DRIFT

A look at the world-map gives the idea that continents and oceans are the principal surface features of the globe. It is also observed that there is a remarkable accumulation of landmasses in the northern hemisphere whereas in the sourthern hemisphere there is appreciable concentration of water bodies. It is also noticed that majority of the continents have southerly tapering ends. Besides, while the north pole is located in the Arctic-ocean,the south pole is located in the continent. Apart from the aforesaid observations, it is also important to notice that continental boundaries on either side of the Atlantic-ocean match with each other. All these observations usually give the impression that the continents might once have been joined but then split apart and moved away from one another as a result of which the present day arrangement of continents and oceans has been developed.

Continental-drift refers to the horizontal movement of the continents on a vast-scale . The idea of moving continents is not new. The notable observations as already mentioned prompted study and speculation by earth-scientists. The similarities in shape of the Atlantic coastlines of South America and Africa were noticed by Francis Bacon in 1620. In 1858 Antonio Snider showed on maps how these continents might have once been joined and then split apart in later times to form the present-day continents and oceans. In the late nineteenth century, a famous Austrian geologist, Edward Suess, suggested that all the southern-hemisphere continents, together with India, were once assembled into a single large landmass which he called Gondwanaland. His idea was based on the similarity of certain geological formations from continent to continent. In 1908, it was F.B. Taylor who propounded the Continental-drift hypothesis, incorporating much other geological evidences. In 1910 and 1912, Alfred Wegener, a German meteorologist, advocated that continents had drifted apart and suggested mechanisms by which this might have occurred.

Taylor's Hypothesis

According to Taylor, long back in the past there were two great landmasses, one at the northern-hemisphere known as Laurasia and the other at the southern-hemisphere known as Gondwanaland. He supposed that in course of time these landmasses (which he pictured as being continuous sheets of sial) started spreading outwards towards the equator, more or less radially from the polar-regions, much as a continental ice-sheet would do. His assumptions of continental-drift were mainly intended to explain the formation of Tertiary mountains.

Taylor concentrated his view on the sudden start of mountain building at the beginning of tertiary in contrast to the quieter period of mid-cretaceous and suggested that this was due to the sudden increase of tidal action of the moon which was captured by the earth in late-cretaceous and the increase in the tidal force was sufficient to increase the rate of rotation of the earth and also dragged the continents away from the pole.

In support of his assumptions he cited the location of most of the tertiary mountains which are found more or less towards the equator side of the continental masses.

Taylor's assumptions, in fact, fail to justify the exact happenings; for example, tidal forces never increase the rate of rotation of the earth; besides, the equatorial movement of continents as assumed by Taylor fail to explain the drifting of South America from Africa. As such, Taylor's hypothesis did not receive much attention.

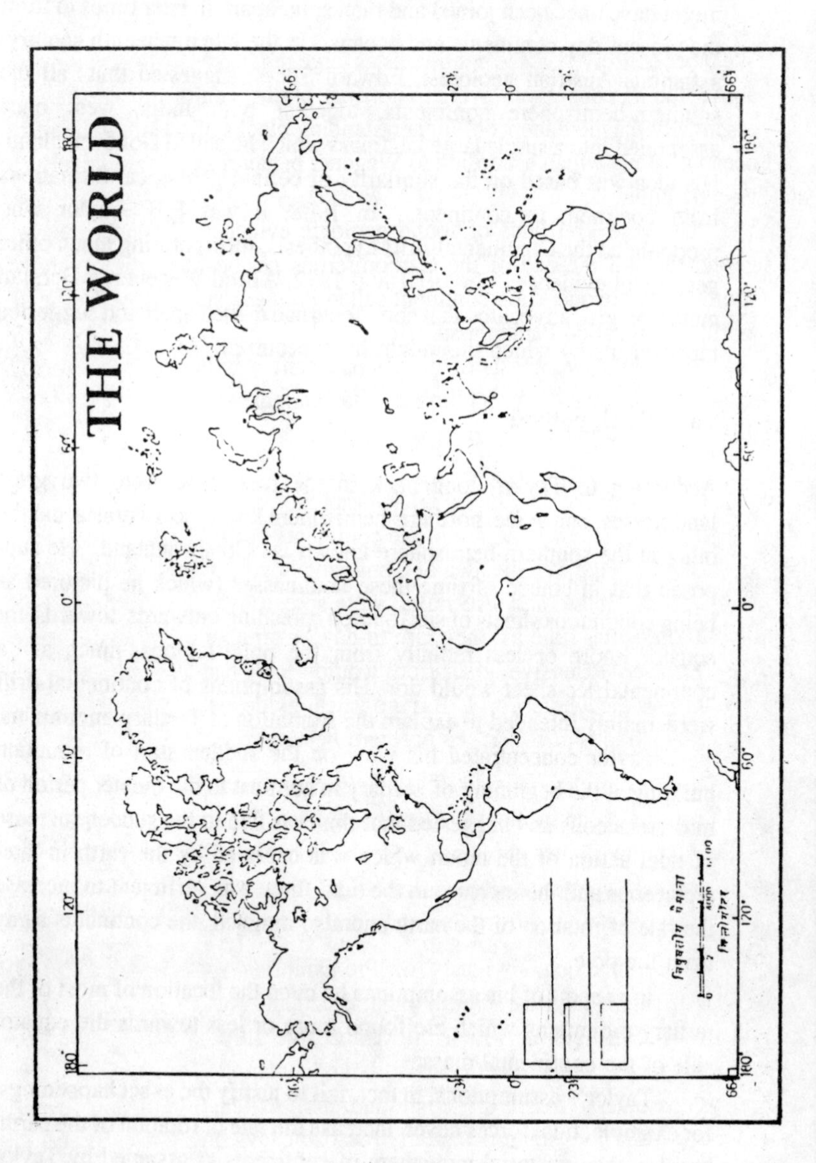

FIG.39.

Wegener's Hypothesis

Wegener conceived of the continents as large thin cakes of sial floating above the viscous sima-layer. The drift of the continents is, thus, essentially the movements of higher-standing blocks of sial through an ocean of sima, which according to Wegener behaves as an excessively viscous liquid.

Using several lines of sound geologic evidence, Wegener suggested that till the end of the Carboniferous period, the present-day continents were one supercontinent called *Pangea*, surrounded by the world-ocean called Panthalassa.

Wegner's hypothesis of continental drift was based on certain palaeontological and palaeoclimatic data available at the beginning of this century. Some of these are as follows:

1. Unmistakable evidences of widespread glaciation towards the palaeozoic era found on the continents of the southern hemisphere support the idea of Continental-drift. If South America, Africa, India and Australia were spread over the earth in the Palaeozoic time as they are to-day, a climate cold enough to produce extensive glaciation would have had to prevail over almost the whole world. But no such evidence of palaeozoic glaciation is found in the northern hemisphere.
2. The occurrence of plant fossil *Glossopteris* of late palaeozoic age in rocks in South America, Africa, India and Australia would be better explained with the assumption that these continents were once joined, although these localities are now widely separated from one another.
3. There is much similarity of Pre-cambrian rocks of Central Africa, Madagascar, South India, Brazil and Australia. Besides, rock sequences in which late palaeozoic tillites are overlain by thick continental sedimentary rocks containing coal beds and Glossopteris fossils are found in India, Africa, South America, Australia and Antarctica. Early Mesozoic lava flows over lie the sedimentary layers. These sequences of rocks in all the above localities are strikingly similar, eventhough they are widely separated now-a-days. But younger rocks in these localities are

quite dissimilar, indicating that the older rock sequences formed together as a single unit while the continents were together but they started splitting apart in early-mesozoic.

4. Wegener collected palaeoclimatic evidences from the sedimentary rocks of each geologic period and found that many ancient climate-belts were in different positions from the present belts. The shift in climate belts through geologic time is related to the phenomenon of *polar-wandering* i.e. the apparent movement of the earth's geographic North and South poles. There may be two alternative explanations about it :—

(a) Poles had actually moved while the continents were immobile; or

(b) Poles had remained stationery while continents had actually moved.

Wegener plotted curves of apparent polar wandering and adhered to the belief that the continents moved and poles were stationary.

5. There is continuity of tectonic-trends of the blocks in countries like Central Africà, Madagascar, Southern India, Brazil and Australia across their present boundaries.

On the basis of the aforesaid geological evidences Wegener had formulated his hypothesis of *Continental-drift.* According to him, all the sialic layer was concentrated in a large continental mass *the Pangea* and was intact till the end of the Carboniferous period. In the late Palaeozoic period, probably during Permian or in the early Mesozoic time the Pangea broke into pieces and the separated continental blocks began to migrate away from each other. Wegener explained the formulation of fold-areas by buckling of the edges of displaced blocks or their collision.

The southern parts of the Pangea broke apart during Mesozoic and the northern in the Tertiary period under the influence of tidal forces due to lunar-solar attraction as well as those generated by the earth's rotation. Some forces caused the drift towards the equator and the other towards the West.

According to the Wegener's picture of the world in Carboniferous time, the south pole occupied a position just off the present coast

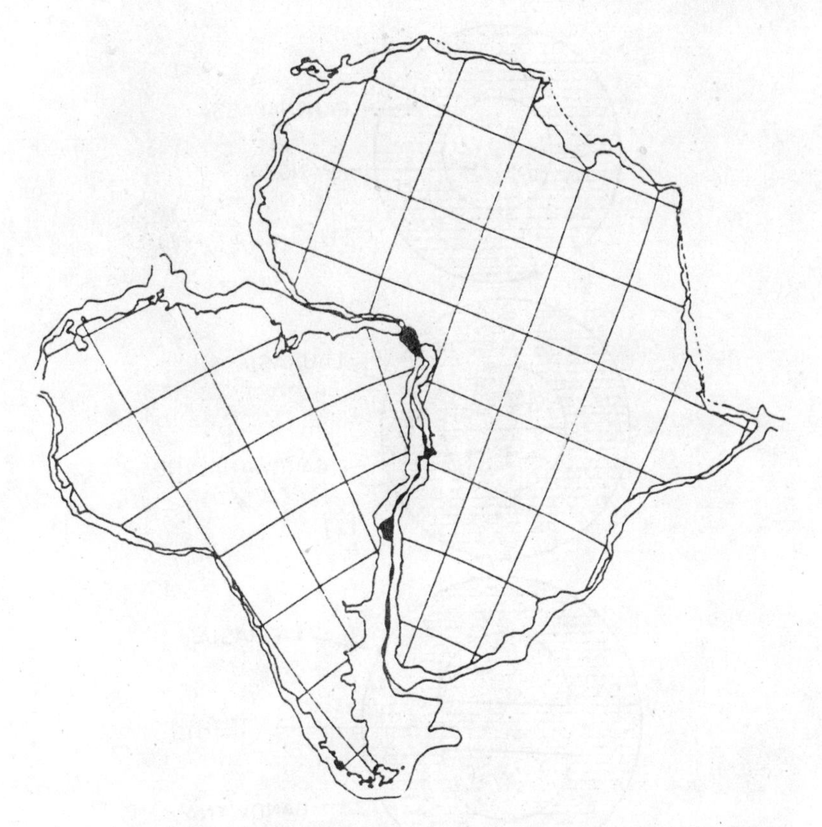

FIG. 40 (A) Diagram showing Zig-saw fit of South Americal and Africa

of South Africa. The drift of the continents away from the poles was termed by Wegener as the *Polflucht* and was believed to be due to the gravitational attraction exerted by the equatorial bulge.

In due course, after Pangea broke apart, the drifting of the detached blocks took place. While the African-block (Gondwanaland) and the Eurasian-block (Laurasia) moved towards the equator, the Americas drifted towards West. North and South America which Wegener somehow pictured as having been attached previous to drifting, rotated about a point in North America and were drawn apart leaving in between only the narrow stringers of Central America and the scattered fragments which now constitute the West Indies. Thus the Atlantic Ocean was created between North and South America in the

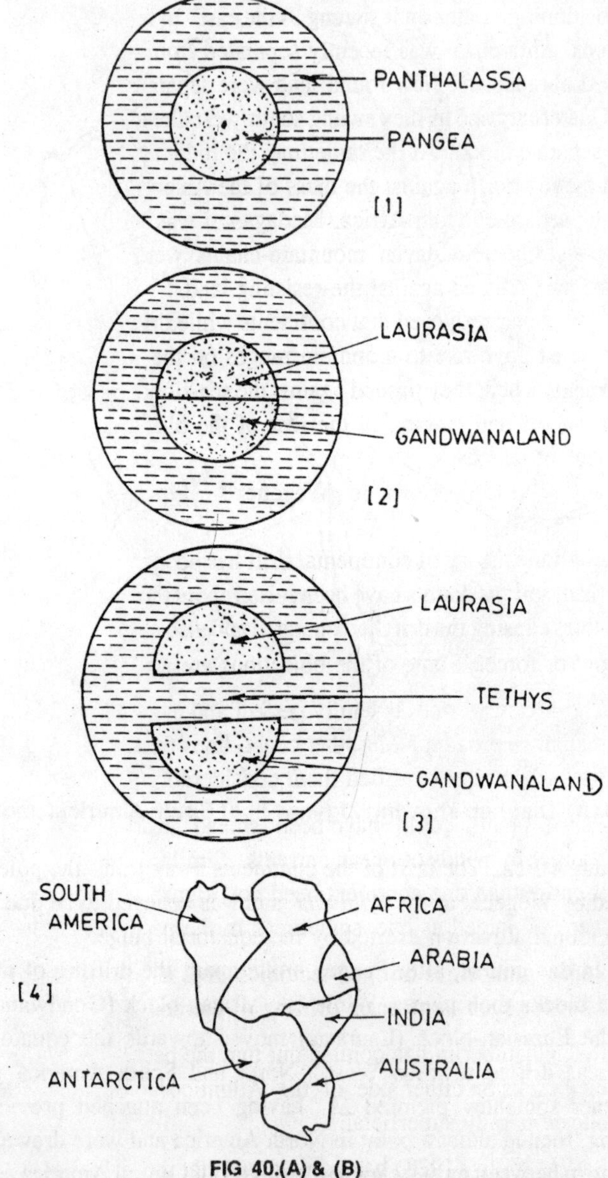

FIG 40.(A) & (B)

SCHEMETIC REASSEMBLED CONTINENTS OF GONDWALAND

West and Europe and Africa in the East. Australia was left behind in the beginning. Later on it swung to the east. In terms of geologic time periods Antarctica was recently separated from South America.

Labrador and New found land were separated from Europe during Quaternary and as they swung south-west Greenland was left behind as a separate block. At the sametime, the Indian part of the Gondwana land moved north against the mass of the Asian main continent, after getting separated from Africa. Madagascar was left behind as a fragment and the Himalayan mountain-chains were created where the Indian part pressed against the resisting land.

Wegener proposed that continents while drifted through the oceanic crust gave rise to mountain ranges on the leading edges of the continents where they pushed against the sea-floor (sima). Landmasses breaking off and lagging in the rear of the blocks moved producing festoons of islands.

Objections to Wegener's Hypothesis

Although the theory of continental drift has been widely discussed and accepted, serious doubts have been raised about the period during which the forces causing the drift had operated and also about the direction and amount of force. Some of the objections are also as follows:

1. The *fit* suggested is hardly perfect and cannot be expected to be real since erosion along the coast lines through long geological ages must have modified them considerably.
2. Fossil plants could have been spread from one continent to another by winds or ocean currents. Similarity of fossils occuring in more than one continent need not signify that the continents were all joined as one, sometimes in the past.
3. Some authorities on geo-tectonics feel it mechanically impossible for such large scale drifting of the continents.
4. R.T. Chamberlin has pointed out that the petrographic analysis of the rocks on either side of the Atlantic shows that their resemblance is only superficial.
5. Bailey Willis (1928) has pointed out that the forces which were great enough to have built the Andes-mountains, would almost certainly have deformed South America to such an extent that any

similarity of its eastern coastlines facing Africa would have been destroyed and the drifting is quite unlikely.

6. Polar-wandering might have been caused by moving poles rather than by moving continents.
7. The mechanism suggested by Wegener for the displacement of continental blocks did not withstand criticism. According to him, tidal forces due to lunar-solar attraction and the forces of the earth's rotation were responsible for the drifting of the continents. But it was difficult for geologists to conceive of a driving force that could move continents. Sir Harold Jeffreys demonstrated that the forces suggested by Wegener to have caused the drift were either ineffective or inadequate for such movements.

In 1937, Wegener's views were developed by the South African geologist Alexander Du Toit in his publication '*Our Wandering Continents*'. Taking into account new palaeoclimatic and palaeontological data, Du Toit advocated that instead of a single Pangea, there had been two major continental masses originally—Laurasia in the northern hemisphere and Gondwanaland in the south-separated by an oceanic area called *Tethys*; a sort of *proto-Mediterranean*. The Alpine-Himalayan mountain chain developed in the place of Tethys by the convergence of the two continental masses.

To explain the horizontal displacement of continental masses Du Toit proposed a different mechanism. According to him, the splitting of the marginal sectors of Gondwanaland was associated with epeirogenetic oscilations of the crust during the sinking and formation of geosynclines. The drifting of the broken-off blocks was associated with an injection of magma and *opening* of continental blocks. But this mechanism, however, has no quantitative basis and was overwhelmingly opposed.

Inspite of vehement opposition from a great number of authorities, the concept of 'Continental drift' was kept alive with enormous efforts in the 1940s and early 1950s, by Arthur Holmes in England, L.C. King in South Africa, S. Warren Carey in Tasmania, etc. With the advent of new palaeo-magnetic studies, sufficient interests were regenerated about continental-drift and this set the stage for modern concepts like *ocean-floor spreading* and *plate-tectonics* which began to emerge in 1960, which lend support to the theory of Continental-drift.

Palaeo-Magnetism

The earth acts as an enormous magnet with its north and south poles lying almost beneath the north-pole and south-pole of the axis of rotation of the earth. It is believed that with the movement of metallic iron in to the core of the earth, the earth acquired a magnetism. Certain rocks record earth's magnetic field in them. These rocks record the strength and direction of the earth's magnetic field at the time of their formation. As an example, the mineral magnetite (Fe_3O_4) exhibits permanent magnetism, so also the mixtures of magnetite in solid solution with titanium oxide. Even, the more highly oxidised mineral form of iron, hematite (Fe_2O_3) also exhibits permanent magnetism. Most rocks have, as a minor constituent, some one or another of these permanently magnetised minerals. The magnetism of old rocks can be measured to determine the direction and strength of the earth's magnetic field in the past. This is known as 'palaeomagnetism', or remnant-magnetism or fossil magnetism.

At the time of cooling of the magma the paramagnetic minerals acquire earth's magnetism and get themselves aligned in the vector of the earth's magnetic-field, thus preserve a record of the earth's magnetic field when the magma solidifies. This is known as thermo-remnant magnetism (TRM). In sedimentary rocks the paramagnetic minerals may be aligned in the earth's magnetic field during their deposition. In case of physical alignment of the particles during the general process of sedimentation, the term *detrital-remnant magnetism* is used.

The study of palaeo-magnetism has made it possible to find out the direction and dip of the earth's magnetic field during different geologic periods. It has been observed that with a few exceptions rocks of a given age throughout the world had the same magnetic polarity. The change from one polarity to the other occurred more or less simultaneously and in a comparatively short time.

The polar-wandering curves drawn, for different continents during a particular geologic period, on the basis of the palaeomagnetic evidences show that they are not parallel or sympathetic. As for example, the analyses indicate that while the Permian lava flows of North-America show a north magnetic pole position in Asia; information obtained fromPermian-lava flows of Europe suggest a different location for the pole in Permian time. Similarly data taken from lava flows from all continents show that each continent is having its own series of north-magnetic pole positions. But, since it is highly unlikely that there were different north-magnetic poles for each continent, it can be inferred from the data that the pole stood still and the continents moved apart and rotated in relation to one another as they diverged.

According to the results obtained from palaeo-magnetic studies the noth pole had at one time been very much farther from India than it is at present. This indicates that there had been a rapid northward drift of India. According to Arthur Holmes Palaeomagnetic work on Deccan Plateau-basalt shows that India referred to the lattitude of Bombay (for easy comparison) drifted northward from 37°S to 13°S during the period of lava extrusion i.e. from Late-cretaceous to the Early-eocene. The present lattitude of Bombay is 19°N. Thus, there is a displacement of about 56°C or in other words, about 5,000 kms in 70 million years i.e.about 7 cm per year. Alongwith this northward drift there is also an anticlockwise rotation of the Indian sub-continent by about 25 to 30°. This resulted in the increase of the tringular gap between Arabia and west coast of India.

These facts give the best proof in favour of the reality of continental-drift.

Continental Drift in the Light of Plate-Tectonics

Closely connected with the conception of the continental drift is the more recent concept of Plate-tectonics. A plate is a large, mobile slab of rock and is part of the earth's surface. The idea underlying the hypothesis of plate-tectonics are associated with the discovery of zones of formation of young oceanic crust (mid-oceanic ridges) and zones of the absorption of the crust (Trenches). The plates constitute the rigid outer shell of the earth called the 'lithosphere' which include the crust of the earth(both oceanic and continental) and the upper portion of the mantle, which overlies the asthenosphere', a zone located at 100 to 150 kms depth and behaves plastically. It is a low-seismic velocity zone.

These plates have thickness of 0 to 10 Kms at the ridges to 100 to 150 Kms elsewhere. They move with velocities of 1-6 cm/year. According to the available data the lithosphere consists of six major plates bounded by zones of spreading, subduction and folding. These are the Pacific, Eurasian, American, Indian, African, and Antarctic.

A number of factors like-oceanic topograhy, gravity, temperature difference between the ridge-crest and the ocean-floor, convection-current etc. are believed to play major roles in the movement of plates. Thus the passive continents move on the spreading oceanic-crust, which supports the theory of Continental-drift.

14

ISLAND ARCS

These are the arcuate chains of volcanic islands which mostly occur adjacent to continental margins. These are the sites of greatest instability found anywhere on the earth and are the most active of the various segments of the crust. Here, processes involving the crust are in action even at the present time. It is now recognised that these are the places where the present orogenic (period of mountain building) and igneous (crystallized magma) activities are taking place.

Most of the island arcs are located in the Pacific Ocean where they almost form a rim around the entire northern, western and southern margins. Others are located in the Atlantic, especially in the Caribbean Sea. The followings are considered to be the modern island-arcs—

(i) New Zealand & Tonga.

(ii) Melanesia.

(iii) Indonesia.

(iv) Phillipines.

(v) Formosa and West-Japan.

(vi) Marianas and East-Japan.

(vii) Kurile and Kamchataka.

(viii) Aleutian and Alaska.

(ix) Central America.*

(x) West-Indies.

(xi) South America.*

(xii) Western Antarctica.

GENERAL FEATURES

1. The islands form arcs which are commonly arcuate-shaped with a length of the order of 1000 kms and width of about 200 to 300 kms. They are invariably concave towards the nearest continent and convex towards the deep ocean-basin.

2. All the arcs are marginal to continents. They are situated just at the contact of the continental crust and the oceanic crust.

3. The island arcs are invariably associated with deep-sea trenches which are commonly found along the ocean-facing side of island arcs. They contain the greatest depths found anywhere in the oceans. These trenches are elongated furrows extending sometimes up to 12600 metres below the sea-level. The trenches are quite asymmetric in nature.

4. The trenches are characterised by large negative gravity anomaly (i.e. the force of gravity over the trench is less than would be predicted for a theoritically homogenous earth). It indicates that there is a mass deficiency in the section of the earth beneath these anomalies. A strong positive anomaly zone is found towards the concave side.

5. They are the sites of active volcanism, which is characteristically andesitic. There is usually noticed a lateral variation in the composition of the volcanic rocks from tholeiitic basalt on the

*The Central and South America are not exactly islands but they have most of the characteristics of arcs.

oceanic side of an island arc system to increasingly alkaline and andesitic lavas on the continental side. The line separating these compositional differences is called the Andesite-line.

6. They are characterised by active seismicity. The belts of greatest volcanic activity of the world run parallel to the arcs. Almost all the earthquakes are very deep-seated and may come from depths as great as 700 km. Shallow quakes are heavily concentrated along the arcs, also. If the positions of the earthquake-epicentres and their depths are plotted, it is found that the deepest quakes are on the continental side and that the more shallow quakes line up in the vicinity of the bottom of the trenches. The line of quake foci commonly dip about 45° or a bit more towards the continents and is generally thought to be a major fault over which ocean basins and continents are moving in relation to one another. They are the sites of active deformation.

7. The island arcs exhibit the petrological features as follows:

Igneous Rocks — Plutonic rocks like granodiorites are abundant in the arcs. Andesitic lava and tuffs mainly constitute the arcs.

Sedimentary Rocks — The sedimentary rocks include gray wackes, eugeosynclinal sediments with ophiolites, organic sediments consisting of limestones etc. in a highly deformed state.

Metamorphic Rocks — High pressure and low temperature metamorphic rocks of greenschist facies are also found present.

8. The largest and youngest batholiths are found in the arcs. But, there is no evidence of the Pre-Cambrian basement of meta morphosed gneisses, which dominate in the shields and which are exposed in the core of the folded mountains.

9. A chain of islands arranged in the shape of an arc, but the curvature may be different in different arcs. Most of the arcs are not directly connected with the continents. They are sepa

rated by seas which in several cases are more than 160 km wide and are relatively shallow.

10. When a double arc occurs, the inner one is volcanic, the outer one non-volcanic. Most of the arcs end where they intersect a second arc.
11. The modern island arcs are of Cenozoic age i.e. less than 65 million years old.

ORIGIN OF ISLAND-ARCS

A number of opinions have been framed regarding the origin of island-arcs, which are conveniently grouped into three categories as follows:

1. Views explaining the origin of island arcs by processes of orogenic deformations. These views have been propounded by Suess, Tokuda, Wegener and Hobbs etc. But, since these views have left much to be explained for many of the features associated with island arcs, they are not commonly accepted.
2. Venning Meinesz, Bijlaard, Hess, Umbgroove, De Sitter have formulated their theories to explain the interrelationship be tween morphologic-cum-other structural features and the inner features like infrastructure, gravity and seismicity anomaly etc. But their views failed to be compatible with physico-mathematical models and as such did not receive much acceptance.
3. The modern theories of Plate-tectonics have been able to explain the formation of island-arcs and is widely accepted at present.

The plate-tectonics interpretation of island arcs is that they are sites of under-thrusting on a large scale, along which the lithosphere of one plate disappears or is consumed as it passes beneath the edge of another plate.

During the movement of the plates sometimes thin, dense oceanic plate margins run into a thick, low-density continental margin. The rocks of the oceanic plate are always denser than the rocks of continents and so the oceanic plate, is flexed downward and forced under the continental crust where it may be wholly or partially melted. The ocean floor is buckled downward, forming deep oceanic trenches or foredeeps.

The earthquake foci associated with any given trench are found to lie in a narrow zone that intersects the surface of the earth in the general vicinity of the trench, and which dips beneath the adjacent continental margin or arc. This downward-sloping seismic zone which may reach upto 600 to 700 kms is termed as Benioff-zones after Hugo Benioff, the geologist who first described their geometry. The fact that there are no earthquakes at depths greater than 700 kms supports the idea that the oceanic plates are melted or sottened by the time they reach this depth. Along the Benioff Zone the descending plate disappears or is consumed by complete or partial melting.

Geologists refer to zones of oceanic plate consumption as subduction zones. The vulcanicity and high heat flow associated with trenches are thought to be associated with the partial melting of the descending oceanic plate including entrained sediments at its upper surface, which return to the surface as volcanoes. Subduction angle determines the arc-trench spacing. If the subduction angle is gentle the arc trench distance is greater, which may be illustrated as follows:

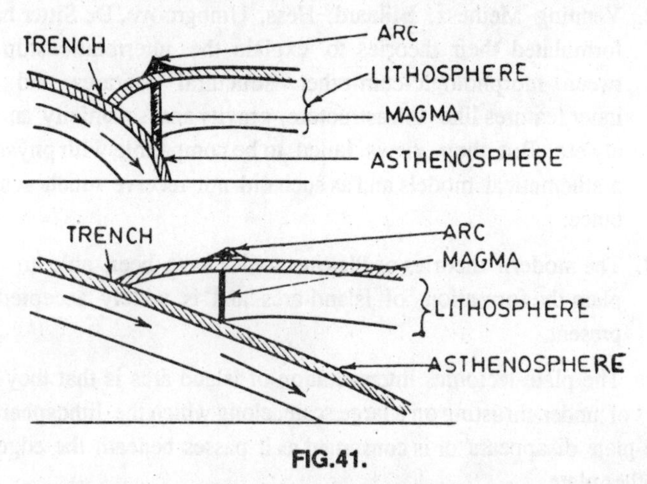

FIG.41.

So far as the lithologic features are concerned, there occurs a complex mixture of deep-water sediments, sedimentary material derived from the oceanic landslides from the continental rise and ocean floor basalts scrapped off the advancing margin of the subducted plate, seaward from the island arcs. This lithological mix is called the *melange*. Besides, it also explains the occurrence of blueschists as well as ophiolites and that of andesitic lava.

A modern example of subduction of oceanic plate directly beneath a continental one is furnished by the collision of Nazca plate and South American plate. The Nazca plate is being forced under the western margin of South America at a rate of 5 cms per year.

Apart from the above, there can also be collision of two plates capped by oceanic crust, in which case also, one plate dives (subducts) under the other. (For details, please see the chapter *Plate-tectonics*).

15

MID-OCEANIC RIDGES

A few years back, the bottom of the ocean was commonly believed to be an extensive flat, featureless plain dotted here and there by isolated volcanoes. But detailed oceanographic work has established the existence of somewhat a linear and comparatively narrow zone that stands at higher level than the ocean-bottoms. These submarine relief features are elongated and fairly continuous. They generally remain submerged beneath the oceanic water and tend to occupy medial positions in each of the main ocean basins. These are commonly known as 'Mid-oceanic Ridges' and they provide evidences of global-tectonics.

By far the largest and the best known ridges is the Mid-Atlantic Ridge, which extends from near Iceland to a point in the South-Atlantic southwest of the cape of Good Hope. The oceanic ridge system stretches nearly 80,000 Kilometres around the world. Apart from the Mid-Atlantic Ridge, the other well-known ridges are the Carlsberg Ridge in the Indian Ocean, the Lomonosov Ridge in the Arctic Ocean and the Pacific-Antarctic Ridge which lies between Antarctica and New Zealand and Australia. In the Indian ocean the ridge is about halfway between Africa and India and extends upto the Red Sea; in the Pacific Ocean the ridge runs along the southern margin of the ocean basin to form a complex pattern off South America that runs northerly into the Gulf of California.

GENERAL CHARACTERISTICS

1. Mid-ocean ridges are typical submarine relief features, They are, in general, submerged beneath oceanic water, however, local crowning above the level of oceanic water gives rise to islands such as Iceland, Mauritius, Laccadive etc.
2. The ridges are fairly continuous with a sinuous pattern of distribution. They are centrally or latero-centrally placed within the oceans.
3. The width of the mid-oceanic ridges fluctuates within the range of 2000 to 3000 kms. The height above the adjacent basins is 2-3 kms and more. Hydrographic work in the North-Atlantic (Tolstoy and Ewing, 1949, Tolstoy 1951) has thrown light upon the topography of the Mid-Atlantic Ridge. Their works indicated that the ridge consists of three rather distinct types of topography:

   (i) There is a high central zone, the Main-Range, which consists of several parallel ridges extending in a general north east-southwest direction. This is also known as 'Crestal Province' and is characterised by block-faulting. The topography usually consists of a central-rift valley, lateral block-mountains and steep scrap surfaces that bound the rift-valley and inwardly slope into it.

   There is a deep valley along the axis of the mid-oceanic ridge, which is a narrow graben bounded by faults. The width of the graben is of the order of 15 kms and its depth below sea-level is around 2500 metres. This valley dissects the ridge in a meridional direction for almost its whole length. Such valleys are called rifts and are considered characteristic elements of mid-oceanic ridges.

   (ii) On the flanks of the Main Range, there is a series of flats, which were designated as the Terraced Zone. Individual terraces vary in width from 2 to 80 kms and collectively are 350 to 450 kms in width.

   (iii) A third zone lies between the terraced zone and the

seafloor plain. It is mountainous and rather distinct from the other two zones. It has individual peaks and is designated as the foot-hills of the Mid-Atlantic Ridge.

4. The mid-oceanic ridges are dissected into separate segments by the so-called transformed deep-fractures, perpendicular to the line of their strike. These faults have caused huge horizontal displacement of the crestal provinces along them. Such faults are called *transform-faults* in the concept of global-plate tectonics, since surface area is neither created nor destroyed along such a fault.

5. Mid-oceanic ridges are characterised by an abnormally high value of terrestrial heat flows. In general the heat flow values are in conformity with the high thermal activity attested by volcanic eruptions at the crestal province. Many active volcanoes are located along ridges. But, the heat flow pattern remains largely unknown.

6. Mid-oceanic ridges are the sites of active seismicity. The earthquakes that occur beneath ridges originate at shallow depth.

7. Mid-oceanic ridges are characterised by a high degree of per meability, expressed by intensive magmatism, which display abundant basaltic volcanism of a low—potassium, olivine-deficient type of lava. Approximately two—thirds of the annual lava eruption on the earth's surface is along the oceanic ridges.

8. So far as the lithologic composition of the mid-oceanic ridge is concerned, the rocks dredged, cored and exposed along the oceanic-ridges are basic igneous rocks and volcanic rocks most of which are basaltic in composition. The samples include such rocks as olivine-gabbro, serpentine, basalt and diabase.

Local acidic volcanics represent the magmatic differentiates of basaltic magma.

Comparatively a thin sedimentary veneer may be found present on the crestal mounts and plateaus.

9. Cross-sectional pictures of the mid-oceanic ridges obtained through geo-physical study shows the presence of basaltic root, about 30 kms deep into the lower mantle for a height of the ridge of the order of 1.6 kms above the ocean botttom. This deep rooting ensures isostatic balance.

10. They are characterised by positive gravity anomaly. Also palaeomagnetic studies on the mid-oceanic ridges reveal that there exists a bilaterally symmetrical arrangement of the magnetic anomalies along the ridge. In other words, the normally and reversely magnetised rocks on one side of the ridge were the mirror image of those on the other side.

11. According to the theories of *Plate-tectonics*, the mid-oceanic ridges mark the divergent-type of plate boundaries.

ORIGIN

Several theories have been proposed to account for the mid-oceanic ridges. According to these theories—

(i) it is a horst;

(ii) it is an anticlinal fold due to lateral compression;

(iii) it is the bottom of the rift which opened as Gondwanaland began to fragment (continental-drift);

(iv) it is produced largely be extrusion of volcanic rocks along the linear openings on the sea-floor;

(v) it is believed to be an orogenic mountain belt similar to those on land;

(vi) it is an outcome of the convection current mechanism. Much convection with in the upper-mantle owing to differential temperature was believed by Holmes. This hypothesis explains the existence of the ridge in two possible ways:

(a) The ridge may be pushed up from below by the rising mantle rock, or

(b) The expansion of the hot mantle-rock may cause the bulge of the ridge to form as a result of the increased volume of the mantle rock.

The high heat flow associated with the ridge crest is plausible, if hot mantle rock is rising beneath it. The active volcanism at the ridge-crest is due to the upward rising of very hot but still solid mantle rock beneath the mid-oceanic ridge, moving from a region of high pressure to a region of low pressure. The drop in pressure lowers the melting point of the molten rock, so that some of the hot-mantle rock melts without the addition of any new heat. The melted mantle rock gives rise to magma near the ridge crest and the magma erupts as basaltic lava.

The rift-valley of the mid-oceanic ridge crest is formed when the rising mantle-rock splits and diverges sideways. As tensional crack opens, there occurs shallow-focus earthquakes.

These assumptions very nicely fit into the structural frame-work of mid-oceanic ridges. But the geo-dynamic picture envisaged in this concept is entirely hypothetical.

All the theories advanced, as already explained, to account for the origin of the mid-oceanic ridges, have been discarded due to their failure in explaining the features associated with the ridges viz. form and alignment of ridges, central rift zone, offset fractures, rock-types comprising the ridges, seismicity of ridges, infra-structure etc. The concept of ocean-floor spreading seems to offer a more logical explanation regarding the formation of mid-oceanic ridges.

Concept of Sea-Floor Spreading and Plate-Tectonics

The corner stone of plate-tectonics is the theory of sea-floor spreading. According to this theory the floor of the ocean is moving or spreading away from ridges, by repetitive magma intrusion that split and spread the older sea-floor, causing them to be moved away from the ridge in a nearly continuous horizontal position. This is also confirmed by the occurrence of progressively older rocks with increasing distance from the ridge. The lack of pelagic sediments at the ridge crest can also be explained by sea-floor spreading. The new floor at the crest is too young to have received a blanket of pelagic sediments. The idea of sea-floor spreading could provide an explanation for aspects of ridge vulcanism, seismicity, and physiography, and for the apparent youthfulness of sea-floor rocks compared with the continental ones.

Plate tectonics incorporates sea-floor spreading as one of its essential parts and makes explicit the idea that the sea-floor and the litho-sphere beneath it are simultaneously being created at ridges. According to it, a diverging boundary on the sea-floor is marked by the crest of the mid-oceanic ridge. It is also believed that upward convection is a result of plate divergence (for details see Plate-tectonics).

16

SEA—FLOOR SPREADING

The topography of the sea or ocean floor is as complex and varied as those on the surface of the land. The earlier idea that the floor of the ocean basins were largely vast, flat to undulating plains with relatively little relief, dull and featureless was proved to be incorrect by the results obtained from the analysis of the outcomes of various geophysical techniques, echo-sounding, coring, bathyscaphe descents etc. It was confirmed that the ocean basins are characterised by the following important features.

— Mid oceanic ridges, along which the formation of new oceanic crust takes place.

— Abyssal plains, which are extremely flat and monotonous in relief occur at depths of 3500 to 6000 metres.

— Isolated peaks rising from the ridge or plains which sometimes are emerged at the ocean surface to form islands. Besides, there are abyssal hills, sea-mounts, guyots etc.

— Deep, narrow oceanic trenches, where the old oceanic crust is destroyed.

The combination of the above discoveries which include the creation of new oceanic crust at the crest of the mid-oceanic ridges and its destruction at the deep oceanic trenches, with the movement of rigid sections in between led to the idea of sea-floor spreading in the early 1960s. Besides, the results of palaeomagnetism and of marine geology and geophysics set the scene for Hess's formulation of the concept of sea-floor spreading. Professor Harry Hess of Princeton University was first to formulate the hypothesis of 'sea-floor spreading' in 1960.

Hess postulated that the major structures of the sea-floor are the surface expression of the convection processes in the mantle. He states that the mid-oceanic ridges are situated over the ascending limbs of convection currents in the earth's mantle, and the oceanic trenches over the descending limbs of the convection current. According to this concept, the sea-floor moves like a conveyor belt away from the crest of the mid-oceanic ridges across the deep ocean basin and then by plunging beneath a continent or island arc get disappeared, since these are the areas where the oceanic crust is largely resorbed by the mantle. Hess also suggested that the rate of movement of the sea-floor is approximately 1-2 cm per year per ridge flank and that the continents, despite their great age and permanency, have been and are being passively carried on the convecting mantle.

Hess has attempted to explain his ideas about the sea-floor spreading through a number of geological facts as mentioned below-

1. *The Mid-Oceanic Ridge* The Mid-Atlantic ridge has thoroughly been studied for this purpose. Along the axial part of the ridge a large depression is stretched. It is bound by deep fractures called a rift valley. The mid-oceanic ridges are connected with intensive earthquakes, volcanism, tectonic movements of various nature and a high heat flow value. The formation of the ridge has been thought to be due to convection-either by the rising mantle rock or by expansion of the hot mantle.

(a) The occurrence of shallow-focus earthquakes along the crest of the mid-oceanic ridge system is readily explained by Hess's model. The rift-valley at the crest of the mid-oceanic ridge is formed as a large tensional crack, when the rising mantle-rock splits and moves in opposite direction on either side of the ridge crest. As the tensional crack opens, earthquakes occur and the rift-valley tends to be filled up by the basaltic lava erupted at the ridge crest. The eruption of basaltic lavas create new oceanic crust.

(b) Hess argued that the anomalously high heat values, associated with the ridge crest is possible because of the rising of the hot mantle-derived material beneath it.

(c) The occurence of active volcanism at the ridge crest has also been explained by Hess. According to him, there is a drop in the pressure on the hot, solid mantle rock while they move upward beneath the mid-oceanic ridge. The decrease in the pressure lowers the melting point of the mantle rock and thus some of the hot mantle-rock undergo melting without any new addition of heat. Thus basaltic magma is formed near the ridge crest from the melted mantle rock and erupts causing volcanism.

(d) The sea-floor has turned out to be geologically very young, not more than 160 million years old which is only 1/30th of the age of the earth; whereas the ancient rock exposed on the land are sometimes three billion years old. This was thought to be due to the generation of new sea-floor at the mid-oceanic ridges that displaces the older sea-floor which finally disappears by subduction down into the trench i.e. at the sites of converging convection cells.

(e) If the recent rates of accumulation and formation of pelagic sediments are extrapolated over the whole of geologic time, it will be seen that the ocean basin as a whole contain a very thin veneer of sediments and quite small number of sea-mounts. Sea-floor spreading also indicates, that since new sea-floor forms constantly at the ridge crest and moves sideways, there is progressive increase in age away from the ridge crest. Accordingly, there is lack of palagic sediments at the ridge crest while it is progressively thicker on the older sea-floor as it moves.

(f) In 1960, Hess calculated that South America and Africa had both moved 2500 Km from the Mid-Atlantic ridge during an interval which he thought was 250 million years. This gives the rate of movements as 10 mm per year (i.e. 1 cm/year). Rates like 1 cm per year are appreciable in human terms.

2. *Trenches* Sea-floor spreading concept also helps in explaining the occurrence of trenches around the margins of the Pacific Ocean and are interpreted as the lines along which the sea-floor spreading towards the continents plunges beneath lower density continental or ocean-floor rocks at the subduction zone to a depth where the rock materials are fused resulting in volcanic and earthquake activity. Since the rocks of the sea-floor are denser than the rocks of the continents, the sea-floor always tend to slide under continents when the two come together.

(a) The strong negetive gravity anomalies over the deep-sea trenches, where the isostatic equilibrium does not operate, is thought to be due to some active force which must be pulling down the bottom of the trenches. According to the hypothesis of sea-floor spreading, this force is the subduction of the sea-floor caused by cooling and skinking of mantle rock.

(b) Relatively low heat flow values, found in oceanic trenches indicate that the rocks beneath the trenches are cooler than normal.

(c) The trenches are the locales of the earthquake foci and associated volcanic activity. The world's belts of greatest seismic activity occur in the trenches and adjacent island arcs. If the positions of the earthquake epicentres and their depths are plotted it will be found that the deepest quakes (700 Km depth) are on the continental side and the more shallow quakes are in the vicinity of the bottom of trenches.

The hypothesis of the sea-floor spreading indicates that the sea-floor material is subducted beneath a continent along what is known as a *Benioff Zone*. The bending of the sea-floor and the friction resulting from the subduction gives rise to earthquake shocks and generates heat which leads to the fusion and melting of upper mantle material combined with ocean-floor lavas and sediments. Localised melting during subduction may create andesitic volcanoes. Andesite was thought to be formed due to the partial melting of the basaltic oceanic crust or of the mantle which undergoes magmatic differentiation and mixes with the rocks through which it passes. This could create a more silica-rich andesitic magma and the composition of the andesite varies with the increase in the depth of the subduction zone and the distance from the trench.

3. *Aseismic Ridges* These ridges are always earthquake free. These are island chains which often have an active volcano at one end and a line of dead and increasingly eroded volcanoes extending from the active one. Away from the active volcano, the island chain gradually does not appear above sea-level, but there is usually a line of submarine sea-months. This gives the sequence of volcanoes being activated, abandoned and eroded.

Hess termed the eroded, flat-topped sea-mounts as *Guyots.* These extinct volcanoes are progressively increasing in age away from the active volcano. An aseismic ridge seems to acquire its alignment as the sea-floor moves over a centre of eruption (i.e. the fixed point of a lava source). If the sea-floor moves over an eruption centre, the volcanoes would gradually be carried away from it and because of cooling they sink as they are moved.

The above explanations although support the hypothesis of ocean-floor spreading, the study of marine magnetic anomalies and deep-sea drilling supports the concept and most of the geologists accept this concept as a theory.

Marine Magnetic Studies

It has been established that all rocks exhibit magnetic properties one of which is a fossil magnetism which is acquired during the formation of the rock. The direction of magnetization is in response to the prevailing magnetic field. The study of fossil magnetism in rocks suggests that the earth's magnetic field had reversed its orientation several times in the last few million years; in other words the position of the earth's magnetic axis had moved relative to its rotational axis. It also indicates that the polarity epochs and events are synchronous in widely spaced parts of the earth.

Two British geologists, Fred Vine and Drummond Matthews, in the year 1963, suggested that as new oceanic crust is formed at, and spread laterally away from, the ridge crest, it gets magnetized in the direction of the earth's magnetic field when it cools below the curie temperature. If the earth's magnetic field reverses its polarity intermittently, then successive strips of oceanic crust paralleling the crest of the ridge will be alternately normally and reversely magnetized. These successive strips will then cause an increase or decrease in the total intensity of ambient magnetic field, thus producing the series of linear anomalies.

The magnetism of normally magnetized strip of oceanic crust adds to the earth's magnetism and therefore a magnetometer carried over such a strip registers a stronger magnetism than average i.e. a positive magnetic anomaly. Similarly, the magnetism of a reversely magnetized strips subtracts from the present magnetic field of earth and therefore a magnetometer carried over such a strip indicates a weaker magnetic field i.e. a negetive magnetic anomaly.

Vine and Matthews found alternating positive and negetive anomalies forming a strip-like pattern parallel to the ridge crest and that the pattern of magnetic anomalies on one side of the mid-oceanic ridge was a mirror image of the pattern on the other side. Besides, the same pattern of magnetic anomalies exists over different parts of the mid-oceanic ridge. They also noticed that the pattern of magnetic anomalies at sea matches those already known from studies of lava flows on the continents.

Vine and Matthews, on the basis of their observations, suggested that Hess's sea-floor spreading combined with a reversing magnetic field would act as gigantic tape recorder which might provide compelling and detailed evidence for sea-floor spreading. Besides, the known times of the magnetic reversals and the known distances of the magnetic strips from the ridge axis gave direct and precise measurements of the rates of sea-floor spreading. According to the sea-floor spreading hypothesis, the age of the crust at any distance is a function of the spreading rate. Since all flows of particular age, irrespective of their geographic locations, indicate the same polarity for the earth's magnetic field, a geomagnetic reversal time-scale for the past three and a half million years was possible to frame. As magnetic reversals have already been dated from the lava flows on land, the anomalies caused by these reversals are also dated and can be used to find out the rate of sea-floor spreading. Once the reversal time is dated for a piece of sea-floor and its distance away from the ridge crest is known, it will be possible to find out the rate of sea-floor spreading. Such measured rates generally range from 1 to 6 cm per year.

Vine-Matthews hypothesis also predicts the age of the sea-floor. The distinctive pattern of the magnetic anomalies enables the marine geologists to find out their age. Most sections of the sea-floor have magnetic anomalies and it is, therefore, possible to predict the age of the sea-floor in any particular region by matching the measured anomaly pattern with the known pattern. Besides, rocks and sediment cores recovered from holes drilled in the sea-floor helps in measuring the correct age of the sea-floor and is widely accepted by geologists to verify the predicted age and in turn the hypothesis of sea-floor spreading and magnetic anomaly. It has been found that there exists a close correspondance between the predicted age and the measured age of the sea-floor.

Apart from the Vine-Matthews hypothesis of magnetic anomalies, which goes to a major extent in support of the concept of sea-floor spreading, the study of the geometry of spreading and the seismicity of the earth gives sound evidences supporting the occurrence of sea-floor spreading, which can be explained as follows:

Fracture Zones and Transform Faults

Fracture zones are long, narrow zones of the ocean floor characterised by irregular topography, fault surface, elongated trough and ridges and extend for thousands of kilometres across the ocean-floor. Magnetic measurements of the sea-floor reveal striking patterns that seem to be offset along the fracture zones. These magnetic data have been interpreted as evidence that the earth's crust has been displaced in an East-West direction by faulting for long distances. J. Tuzo Wilson of the University of Toronto in 1965 developed the concept of transform faults, which are strike-slip faults and the mid-oceanic ridge is offset along the fracture zone by strike-slip motion.

As sea-floor spreads away from the ridge crest, blocks of rock move in opposite directions only on that section of the fracture-zone between the two segments of the ridge crest. Earthquakes, therefore, occur only in this section of the fracture zone. The direction of motion of the rocks on either side of the fracture zone provides further support for the concept of sea-floor spreading.

As explained above, these evidences have been accumulated to show that the crust is spreading apart along the rift and as spreading occurs, basaltic lava rises from beneath the rift,solidifying andforming new oceanic crust.

17

PLATE—TECTONICS

As we know, the crust of the earth consists essentially of about 35 kms thick layer of solid rock matter which varies in thickness from about 5 km in the ocenic areas to even such thickness as 70 to 80 kms in the mountainous regions of the Alps and the Himalayas. That the crust is not fully rigid, but has been repeatedly deformed in the geologic past and is subjected to movements even at the present time is confirmed by earthquakes frequent on the ocean-floors and rarer in continents, volcanism, folding and faulting of large expenses of rock strata and recent elevation and depression of coastal areas.

Evidences derived chiefly from palaeomagnetic studies and those of earthquake waves show that within the upper mantle there is a soft layer which behaves plastically because of increased temperature and pressure. This layer of the upper mantle is known as 'asthenosphere' and is approximately 150 kms thick. The crust of the earth (oceanic and continental) together with the uppermost portion of the mantle which overlies the asthenosphere constitute the lithosphere. The rigid lithosphere is capable of moving bodily over the asthenosphere and is disjointed into large segments or blocks extensively by faults or thrusts. These block are known as 'lithospheric plates' which are in motion relative to each other.

The 'concept of Plate-tectonics' involves a world-wide net-work of moving lithospheric plates. The concept was formulated by the American scientists Hess and Dietz. The main ideas underlying the hypothesis of 'plate-tectonics' are the out-come of the study of the structure of the ocean-floor and the discovery of zones of the formation of young oceanic-crust on mid-oceanic ridges and zones of the absorption of the crust in trenches. This concept takes into account the satisfactory parts of the hypotheses of 'Continental-drift' and 'Ocean-floor spreading'.

The basic idea of plate-tectonics, as has been stated in the above paragraph, involves the movement of lithospheric plates and the geologic activities associated with them. Accordingly, the total system of plate motions is commonly referred to as 'plate-tectonics'.

IMPORTANT FEATURES OF PLATE-TECTONICS

1. A plate is a large, rigid slab of rock which moves slowly over the asthenosphere.
2. The thickness of a plate is from '0 to 10 Kms at the ridges and 100 to 150 kms elsewhere.
3. The plates are of continental dimensions.
4. So far as the number of lithospheric plates (of which the earth's surface is considered to be made up) is concerned, there exists a divergence of opinions among the prominent geologists in this field. While W.J. Morgan believes that the earth is composed of 20 lithospheric plates, Le Pichon throught it to be composed of six major plates and the smaller plates were incorporated into these chosen six. This number is commonly accepted by most of the workers in the field. These major plates are:

(a) the Pacific plate.

(b) the American plate.

(c) the African plate.

(d) the Eurasian plate.

(e) the Indian, plate, and

(f) the Antarctic plate.

According to the U.S. Geological Survey, there are eight large lithospheric plates and a few dozen smaller plates which constitute the crust of the earth.

5. The plates are continuously in motion both with respect to each other and to the earth's axis of rotation. Each plate is capable of moving independently of its surrounding plates.
6. The plates may contain continental as well as oceanic surface. Of the six major plates, five contain continents or parts of the continents in the lithosphere. Only the Pacific-plate is made up mostly of sea-floor. Some of the smaller plates may be entirely continental but all the large plates contain some sea-floor.
7. Virtually all seismicity, volcanicity and tectonic activity is localized around plate margins and is associated with differential motion between adjacent plates.
8. Plate interiors generally are devoid of earthquakes, volcanoes, young mountain ranges and other significant geologic activities.
9. These plates are small and large, separated by faults and thrusts, lying mostly across ridges or parallel to borders (trenches).
10. Plates move with velocities ranging from 1 to 6 cm per year. Plates move away from one another, past one another or towards one another.

    Where two plates diverge, we find extensional features, typically the oceanic-ridges, symmetrical about the vertical axis. Where two plates converge and one is thrust beneath the other, we find the island arcs—the huge assymmetric features which are the sites of greatest earthquakes, explosive volcanism, great topographic relief and many other distinctive features.

    Where two plates slide past each other, there occurs *transcurrent-faults* i.e. the large strike-slip faults joining segments of ocean-ridges or arcs.
11. Plate boundary is the surface trace of the zone of motion between two plates. While plates are located by outlining the

inactive regions of a map, plate boundaries are located and defind by mapping narrow bands of tectonic activities such as belts of earthquakes and volcanism etc.

Depending upon whether plates move away from each other, move towards each other or move past each other, plate boundaries are classified into three types as Constructive, Destructive and Conservative plate boundary, respectively.

12. Plate margin is the marginal part of a particular plate. Two plate margins meet at a common plate boundary. Where three plate boundaries meet, it is termed as a Triple-Junctions.

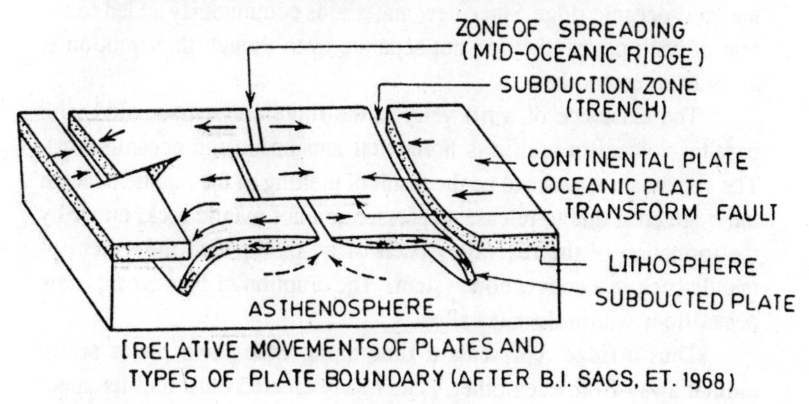

[RELATIVE MOVEMENTS OF PLATES AND TYPES OF PLATE BOUNDARY (AFTER B.I. SACS, ET. 1968)

FIG.42.

PLATE BOUNDARIES

The plate boundaries are the sites of intense geologic activities which are mainly due to the movement of plates. As has already been indicated three types of plate boundries have been recognised on the basis of the movements of the plates such as Constructive plate boundary, Destructive boundary and Conservative boundary.

(i) *Constructive boundaries* These are also known as *diverging plate boundaries.* This is a zone along which two plates are in motion away from each other as a result of which a fissure develops, allowing hot, molten rock materials to well up from the mantle and to form new plate material as it congeals. The fissure represents the zone of spreading and since new crust is created by the upwelling of materials from the mantle, this type of plate boundary is known as a Constructive or divergent margin.

The divergence of plates may take place in the middle of an ocean or in the middle of a continent.

Oceanic Divergence

A diverging plate boundary on the ocean-floor is marked by the crest of the mid-oceanic ridge. The lithospheric plates diverge at the crest of the mid-oceanic ridge. Since new material is continuously added to the rear of each plate, they do not separate even though their motion is away from each other.

The existence of a rift vally at the ridge crest, associated with basaltic volcanism confirms horizontal tension on the oceanic crust. The volcanism appears to be the result of melting of the mantle beneath the ridge crest due to release of pressure on hot mantle rock, either by the formation of the rift valley itself or by the upward movement of mantle-rock in a convection system. The eruption of lava creates new ocean-floor within the rift valley.

Thus a ridge represents a zone along which two plates are in motion away from each other. The newly created crust and its upper mantle is effectively welded to the plates trailing edges. Such boundaries at which the net effect of the plate motion is to create new surface area are termed as *Sources*.

Continental Divergence

This is also known as continental rupturing. In the initial stages of continental rupturing, a column of heated mantle rock begins to rise from deep beneath the continental plate. This causes elevation of the continent. This column of hot mantle rock is called a mantle plume. Due to elevation the crust becomes thinner above the uplift and the tensional forces cause the plate to fracture forming a rift valley along a central graben. As a result, there occurs basaltic volcanism along the fractures within the rift valley.

The fracture system gradually spreads across the entire plate dividing it into two plates. As divergence continues, the continental

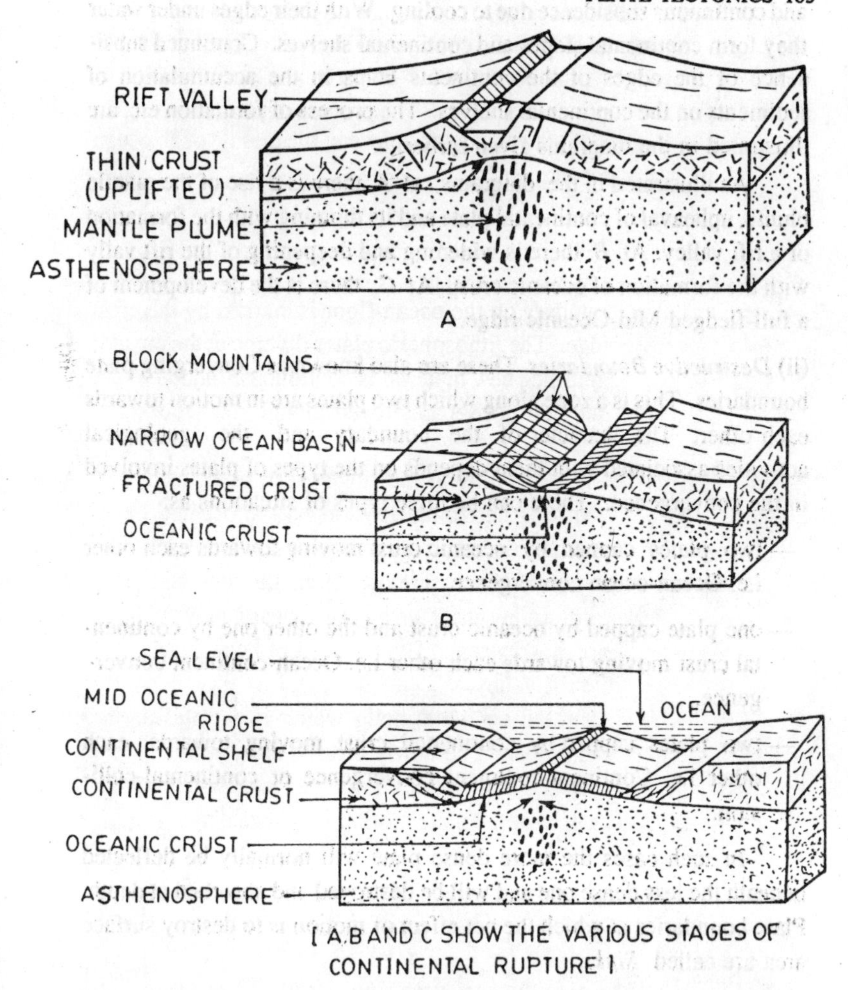

[A,B AND C SHOW THE VARIOUS STAGES OF CONTINENTAL RUPTURE]

FIG.43.

crust on the upper part of the plate distinctly separate with the widening and deepening of the rift valley, admitting ocean water in to the linear basin which forms a narrow ocean between the diverging pieces of the original continent. Continued basaltic volcanism begins to build true oceanic crust between the two newly formed continents and gradually a full fledged mid-oceanic ridge develops in the widening ocean basin.

The trailing edges of the continents are lowered by both erosion and continuous subsidence due to cooling. With their edges under water they form continental slopes and continental shelves. Continued subsidence of the edges of the continents helps in the accumulation of sediments on the continental shelves. The process of formation etc. are illustrated in the diagrams given earlier.

As illustrated in the diagrams, at *A* there is a rise of the mantle plume, upheaval of continental plate and its thinning with the formation of a rift valley. At *B* there is widening and deepening of the rift vally with the formation of oceanic crust. At *C*, there is the development of a full-fledged Mid-Oceanic ridge.

(ii) *Destructive Boundaries* These are also known as Converging plate boundaries. This is a zone along which two plates are in motion towards each other. The character of the boundary and the geological activities associated with them depends on the types of plates involved in the convergence. There can be three types of situations as:

— two plates capped by oceanic crust moving towards each other i.e. Ocean-ocean convergence.

— one plate capped by oceanic crust and the other one by continental crust moving towards each other i.e. Ocean-continent convergence.

— two plates capped by continental crust moving towards each other i.e. Continent-continent convergence or continental-collision.

In such cases the more dense plate will normally be deflected beneath the less dense one and will be destroyed and absorbed at depth. Plate boundaries at which the net effect of motion is to destroy surface area are called *Sinks*.

Ocean-Ocean Convergence

In such cases two oceanic crusts converge and one plate bends downward beneath the other. This phenomenon is called *subduction*.

Subduction of the plate forming the oceanic crust is the cause of the *trench* which is formed immediately over the zone of subduction of the descending plate. At depths between 150 to 200 kms melting of the descending plate takes place due to frictional heat and higher geothermal heat. In the beginning tholeitic lava comes out from the oceanic-crustal melt. Later on andesitic lava predominates (perhaps due to magmatic differentiation). Along with the down bent plate, part of the trench sediments and some water is also carried down to the hotter environment. With the rise in temperature and pressure in depth this water gets released and rise into the overlying plate causing partial melting of rock by lowering its melting point. Even if the original magma is not andesitic, a part of of it may differentiate into andesite.

Being less dense than the surrounding rock, the magma bodies rise slowly to the surface forming a curved line of volcanoes which form a string of volcanic islands (parallel to the already formed oceanic trench) known as *Island arc*. Between the island arc and the continent lies a marginal ocean basin. The eruption of magma takes place, where the top of the downbent lithospheric plate comes in contact with the asthenosphere. Accordingly, the horizontal distance between the island arc and the ocean-trench varies. This distance is determined by the angle of subduction. Steeper the subduction angle shorter is the distance between the arc and the trench. Similarly when the subduction angle is gentle the arc-trench spacing also becomes great.

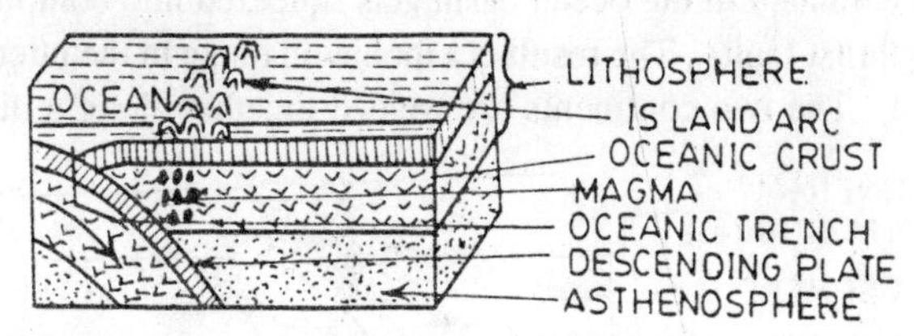

FORMATION OF ISLAND ARC AND DEEP SEA TRENCH]

FIG.44.

Ocean-Continent Convergence

In such cases a plate containing the oceanic crust is subducted under a plate containing the continental crust. The continental crust being less dense is comparatively buoyant and therefore when an oceanic plate converges against a continental plate, it is forced to bend down beneath the latter. The magma that rises from the subduction zone forms a *Volcanic-arc* within the continental crust rather than an island arc at sea. The volcanic-arcs are due to igneous activity on continents and are mainly composed of andesitic lava, with the melting and disappearance of the descending oceanic plate in the hot mantle. But the continental crust remains intact and with the upward rising of hot magma from the subduction zone becomes thicken (i.e. by addition of rising igneous rocks). A young mountain range is formed due to the isostatic uplift of the thickened crust. Sometimes, the edge of the continent is deformed during subduction resulting in a young mountain range.

Continent-Continent Convergence

This phenomenon is also known as continental collision. Here two continents move towards each other. Initially there exists an ocean floor in between the two continents. With the two continents coming closer the ocean basin gets reduced in width and the ocean floor gets subducted under one continent. In the beginning stages the edge of one continent will have a volcanic arc and all the other features of ocean-continent convergence (as already explained).

Continued subduction of the ocean-floor and the narrowing of the ocean basin eventually bring the continents into a collision. The ocean basin gets disappeared completely and the sediment mass which had earlier accumulated in the ocean basin gets squeezed into complicated folds with thrust faults. The result is a mountain range in the interior of a continent. The two continents are welded together along a dipping

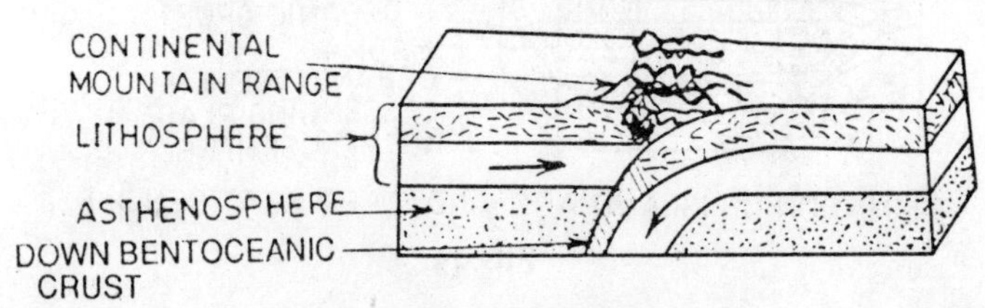

[CONTINENTAL COLLISION]

FIG.45.

suture zone. This zone marks the earlier site of subduction and the mountain range is called a suture. The process of its formation is known as Continental-suturing.

The Himalaya Mountains are thought to have been formed in this process; due to the collision of the Indian plate against the Eurasian plate. The Tethys sea disappeared and the Himalayas came into existence in its place with the collision.

(iii) *Conservative Boundaries* These are the boundaries where two plates slide past one another along a single fault or a group of parallel faults. The motion of the plate is strike-slip along the fault, resulting what is called a transcurrent or transform fault. These are accordingly known as *Transform boundaries.* These are also known as shear margins and the plates neither gain nor lose surface areas. In such cases, little interaction takes place between the plates on either side of the boundary.

Surface Motion of the Plates

Plates move over the surface of the earth, each describing a circular path round its pole of rotation. The formation of new crust in mid-oceanic ridges is accompanied with an absorption of lithospheric plates in the zones of deep-ocean-trenches. Thus, there is continuous movement of lithospheric plates.

'Euler's theorem' is a geometrical concept for measuring the motion of the plates. It shows that the displacement of a plate from one position to the other on the surface of a sphere is a simple rotation of a plate about a choosen axis that passes through the centre of the sphere. All the points on the plate travel along small circular paths from their initial to final position. Plate boundaries parallel to a small circle are conservative in character and other plate boundaries are either constructive or destructive in character.

To understand the mechanisms of the movement of plates, it is necessary to know the physical characters of the lithosphere. Some of the important characteristics are as follows:

(i) *Plate size* Plates vary in size and change in size also. The change in plate-size is evident from the widening of the Atlantic ocean (due to the moving away of the continents from the mid-oceanic ridge) whereas the Pacific-ocean gets smaller (due to movement of continents from all sides).

(ii) *Thermal property* Newly created hot lithospheric plate in the process of moving away from the accreting plate boundary cools down in accordance with an exponential law through flow of heat at its surface.

(iii) *Mechanical properties* At the mid-oceanic ridges the oceanic lithosphere is very weak due to its high temperature of formation than the normal oceanic or continental lithospheric plate.

Continent bearing plates are easier to deform than ocean-bearing plates. The plates are somewhat elastic and bends under super-crustal load.

Causes of Plate Motions

The authors of the theory of plate-tectonics have proposed several different causes for the plate-motions. Some of the important causes are as follows:

1. *Formation of oceanic crust* As it has already been explained new crust is formed in zones of spreading. Since this crustal for mation is a continuous process, the new crust pushes the earlier ones from the ridge crest and causes spreading of the lithospheric plate. This process is believed to cause plate motion.
2. *Oceanic-topography* As the mid-oceanic ridges, from where new lithospheric plates originate, are normally 2 to 4 kms above the level of the ocean floor and slope away more or less symmetrically from the crest, it is thought to be responsible for plate motion.
3. *Rate of Motion* The formation of new crust at the mid-oceanic ridge is accompanied with an absorption of the lithospheric plate at the deep oceanic trench. It is believed that since spreading of plates at ridge-crest occurs at rates ranging from 1 to 6 cms per year but are consumed at the rate of 5 to 15 cms per year at the oceanic trenches the plates must move to maintain a balance between the two.
4. *Temperature difference* The plate being subducted is characterised by low heat flow indicating that the oceanic plate cools, continuously with increasing distance from the ridge-crest as they move from the mid-oceanic ridge to oceanic-trenches. The cooling of the plate increases its density and it becomes heavy

enough to sink back into the mantle. Thus, the difference in the heat flow values seem to be responsible for plate motion.

5. *Gravity difference* Mid-oceanic ridges are almost close to isostatic equilibrium but trenches are marked by very strong negetive gravity anomalies. This gravitational difference is believed to cause plate motion.
6. *Mantle Convection* Convection—current condition in the mantle zone seems to cause plate motion. This is thought to be the slow displacement or convection of mantle-material under the action of the temperature difference between its floor and roof. The velocity of the convectional currents, according to D. Tozfer (1965) is within the range of 1-3 cm per year. The flow of mantle material acts as a kind of fluid conveyor. The diverging current drags the lithospheric plates along the direction of their flow. It is believed that between the rising convection currents (i.e. the mid-oceanic ridges) and the sinking currents (i.e. the subduction zone) the horizontal motion of the convection current exerts a dragging force on the lithospheric plates and causes the plate motion.

Importance of Plate-Tectonic Theory

Plate tectonics has become a unifying theory of geology and geologists have found it as a convenient framework to explain several geological phenomena such as:

(a) Origin and distribution of earthquakes.

(b) Origin of mountain ranges and their distribution.

(c) Formation of island arcs and oceanic trenches.

(d) Distribution of volcanoes.

(e) Continental drift.

(f) Formation of mid-oceanic ridges and rift valleys.

(g) Ocean-floor spreading, etc.

Objections to the Concept

A number of objections have been raised to the theory of plate-tectonics. Some of them are as follows:

(i) There are weak evidences of thermal convection in the mantle, which is usually considered as the driving force for plate motion. The existence of horizontal seismic separation boundaries in the mantle is the evidence of changes in the physical composition of its different layers and consequently of the absence of radial convection currents.

It is being assumed at present that the driving force in the displacement of lithospheric plates is not the thermal but the thermogravitational convection.

(ii) The assumption of pushing down of solid lithosphere into the mantle to a depth as much as 700 kms is not adequately substantiated.

(iii) The continental drift hypothesis has not so far produced any tenable interpretation concerning the planetary net-work of long-living deep faults common to continents and oceans.

(iv) The finding of Precambrian and Cambrian rocks near the crest of the Mid-Atlantic ridge contradicts the theory of plate-tectonics. According to the plate tectonic theory the rocks of the sea-floor cannot be more than five or ten million years old.

However, the plate tectonic theory explains a number of geological phenomena quite convincingly and it is considered to be a revolution in earth science.

18

WEATHERING

Rock-weathering is a phenomenon of the interface between the atmosphere and lithosphere. Rocks which are exposed to the surface are brought to a new environment that is quite different from those under which the rocks were formed and they are acted upon by various natural agencies as a result of which the rocks undergo several changes. Sometimes the rock masses are mechanically disintegrated with the physical forces associated with the natural agencies and sometimes they are decomposed and altered due to chemical reactions. The general outcome is a layer of degraded rock materials resting upon the unaltered rock. Generally the term *weathering* is applied to the combined action of all processes causing rocks to be disintegrated physically and decomposed chemically because of their exposure at or near the earth's surface. In general, weathering occurs, where rocks and minerals come in contact with the atmosphere, surficial water and organic life under conditions that are normal to the surface of the earth.

Weathering is the initial stage in the process of denudation and is considered as the static part of the general process of erosion. An essential feature of the process is that it affects rocks insitu and does not involve transport of the degraded rock materials. The products of rock-weathering, therefore, tend to accumulate as a soft-surface layer, to form a *mantle of waste* or *regolith.* The regolith grades downward into the solid, unaltered rock, which is commonly known as *bed-rock.* Weathering helps erosion to a considerable extent.

FACTORS AFFECTING WEATHERING

A number of natural factors affect weathering in various ways and control the degree of weathering to a major extent. Some of the major factors that affect weathering are as follows

(a) Climate,

(b) Topography,

(c) Structure, texture and mineral composition of rocks,

(d) Vegetation cover,

(e) Time,

Climate Climate is more generally defined as the sum-total of the meteorological elements like temperature, moisture including both humidity and precipitation, atmospheric-pressure, wind etc. Climate determines whether mechanical disintegration or chemical decomposition will predominate and the speed with which these processes will operate.

The climates of various parts of the world are mainly classified on the basis of temperature as well as pressure.

According to the temperature prevailing in a region, it is termed either as:

(i) *Tropical* Where the temperature is uniform through out the year and there is no winter.

(ii) *Temperate* where temperature is highly variable and there is summer and winter.

(iii) *Polar* where temperature is always low mostly below 10°C.

On the basis of precipitation i.e. rainfall the climate of a region may be termed as:

(i) *Arid climate* Where there prevails high temperature and low rainfall.

(ii) *Semi-arid climate* Where the annual rainfall is 25 to 50 cm.

(iii) *Semi-humid climate* Where the annual rainfall is between 50 to 105 cm.

(iv) *Humid climate* Where the annual rainfall is between 105 to 210 cms.

It is commonly observed that physical disintegration predominates in the arid or tropical climatic conditions i.e. in the drier, higher and colder regions of the world. Similarly, rock decomposition is more active in humid climates i.e. in moist, warm and low-lying areas.

Importance of Temperature

The type and degree of weathering depends much on the range of diurnal and seasonal temperature changes. A considerable range of temperature subjects the surface layers of the exposed rocks to expansion and contraction. The thermal expansion and contraction of the rocks with the rising and falling temperatures are expected to cause mechanical destruction of the rocks since the minerals constituting the rocks have different co-efficients of expansion. Besides, alternate freezing and thawing, due to temperature fluctuation also cause shattering of rocks.

Apart from the role played by temperature in breaking down rocks mechanically, temperature also affects the rate of chemical reaction. Rock decomposition is promoted by warmth and humidity.

Importance of Rainfall

Humidity is important in both physical disintegration and chemical decomposition of rocks. As we know, in the polar regions as well as the mountain regions (mostly above snow-line) the intermittent freezing of water present in the porespaces and fissures of rocks split the rock apart and thus cause their disintegration.

Water is an active chemical agent due to the fact that it always dissociates partly into ions of hydrogen (H+) and the hydroxylions

(OH-). Many of the reactions are controlled by the hydrogen ion concentration. It affects the solubility of substances and accordingly affects the chemcial decomposition of rocks, thereby eliminating all the unstable minerals present in the rocks.

Topography

It also controls the degree of weathering through the local relief. Relief, as we know, is the gradient of an area; or in other words, it is the angle of inclination of two specific points divided by the horizontal distance. It is , therefore, the erosive power as well as the transporting power of the streams etc. increases as relief becomes high. Thus, prolonged weathering is marked in the area of low-relief. The most suitable condition for weathering is the existence of an average hill-country topography that ensures percolation of meteoric precipitation down to the water-table.

Structure, Texture and Mineral Composition of Rocks

Structural features of the rock affects the degree of weathering to a major extent. The presence of joints, cracks and fissures in the igneous rocks, the beddings of sedimentary rocks as well as foliations of some metamorphic rocks, facilitates maximum weathering. These features help both mechanical and chemical weathering processes to operate effectively, for instance, water when freezes in the fissures and pores of rocks, exerts enormous pressure on the walls of fissures and make them disintegrate into fragments.

Texture of a rock determines its hardness and strength. It is commonly noticed that fine-grained rocks are most susceptible to weathering, in contrast to their coarse grained counterpart. Besides, it is also observed that harder the rocks less is the degree of weathering and that it is maximum in soft rocks.

The major factors affecting the degree of weathering is the mineralogical and chemical composition of the rocks. As we know, different silicate minerals of which most of the rocks are composed have different stability. The stability order has been represented in the following way:

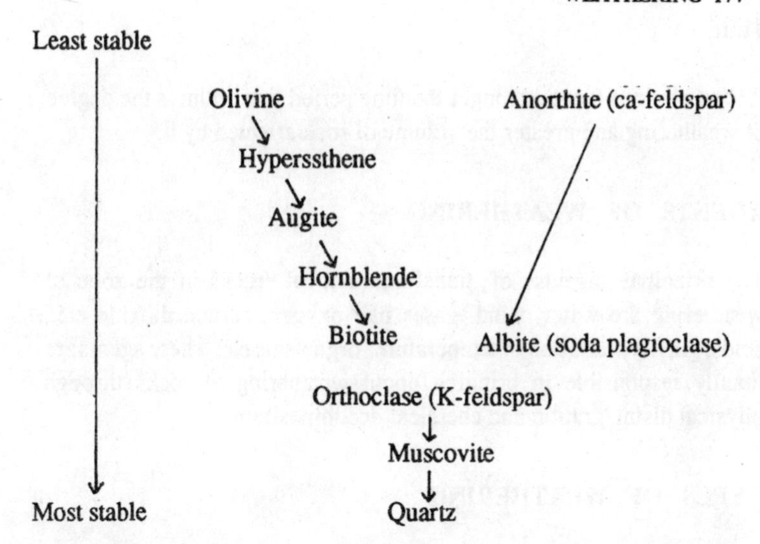

of all the rock-forming minerals, quartz is the most inert oxide and persists under all conditions of temperature and pressure, below that of its formation. Muscovite is a close follower and feldspar comes next. Alkali-feldspars stay longer than the plagioclases. Ferromagnesian silicates, as a group, are less stable than feldspars; biotite being more stable than the amphiboles which are stabler than the pyroxenes and the olivines. It is mostly due to the above fact that basic and ultrabasic rocks are easily weathered in a short duration of time, whereas granites need prolonged weathering; or in other words, basic rocks are more susceptible to chemical decomposition than the acid rocks. Thus, the nature of the pre-existing rocks also controls the degree of weathering.

Vegetation Cover

It is commonly observed that surfaces covered with vegetation are protected to some extent whereas bare-surfaces are weathered to a greater extent. Besides, plant-roots growing between joints and fissures of rocks exert an expansive force tending to widen those openings and thus cause the rocks to disintegrate. Apart from this, there emerges a vast amount of organic acids (humic acids) due to the decay of plants, which help in the decomposition of rocks.

Time

It is an obvious fact that longer the time period maximum is the degree of weathering and greater the volume of rock affected by it.

AGENTS OF WEATHERING

The principal agents of transformation of rocks in the zone of weathering are water, wind, gases like oxygen, carbon dioxide etc., acids, gravity, variation of temperature, organisms etc. These agents are mainly responsible in bringing about weathering of rocks through physical disintegration and chemical decomposition.

TYPES OF WEATHERING

Three distinct forms of weathering have been identified as follows:

1. Physical weathering or Mechanical processes of disintegration;
2. Chemical weathering or Chemical processes of rock-decomposition, and
3. Biological weathering or processes of disintegration and decomposition associated with the activities of organisms.

PHYSICAL WEATHERING

This process refers to the mechanical disintegration of rocks, in which the mineralogical composition of the rock is not affected. It manifests itself as a reduction of the material to smaller and smaller pieces. In simple words, physical weathering is a process of fragmentation of rock due to some physical forces associated with the factors like fluctuations in temperature, change in the pressure, growth of crystals, freezing of water, frost action etc.

(a) *Fluctuation in temperature* Since rock is a poor conductor of heat, it is generally the outer shell of the rock which is subjected to the diurnal changes of temperature i.e. the outer shell of the rock only expands and contracts with the rising and falling of temperature, whereas the inner part of the rock is relatively insulated from temperature changes. It is, therefore, believed that periodic expansion and contraction of rocks leads to the formation of cracks parallel to the heated surface and later to the flaking off of the upper layer. The process of scaly peeling off of the rocks is known as *Exfoliation* or *Desquamation*.

But, laboratory experiments have shown that a small quantity of water and some-degree of chemical activity are necessary before rock flaking will occur. This is also known as 'mass-exfoliation'. The intense heat of forest and bush fires is known to cause rapid flaking and scaling of rocks. Most igneous and metamorphic rocks are polymineralic i.e. composed of several minerals. These minerals have different coefficients of thermal expansion and it causes differential expansion of minerals, thus gives rise to minute internal fracturing. Even in monomineral-lic-rocks the liner co-efficient of expansion of mineral differs from one direction parallel to the crystallographic axis to the other. Thus, monomineralic-rocks also disintegrate due to temperature changes.

Rocks composed of different coloured minerals also undergo differential expansion. This is due to the fact that dark minerals are more strongly heated than the lighter ones. The difference in their volumetric expansion may also lead to the development of cracks and gradual disintegration of the rock. This process is also known as 'granular'-disintegration'.

The weathering due to fluctuation in temperature is termed as 'Thermal Weathering'. It is observed in almost all the climatic zones but is more intense in regions characterised by sharp temperature fluctuations, dry air, absence or poorly developed vegetation cover etc.

(b) *Change in the pressure on rocks* As we know, most of the igneous and metamorphic rocks are formed at great depths under conditions of high temperature and pressure which are very different from those found at the surface of the earth. Besides, with a mean density of 2.7 gm/cm³, the confining pressures exerted on a deeply burried block of rock by the column of overlying material are enormous. As a result of removal of the overlying rocks by denudation, the pressure on the rocks beneath them also diminishes and the block may adjust to this unloading by upward expansion resulting in the development of a closely spaced joint system and fractures parallel to the surface topography. Sheets between the fractures are detached from the main mass of rock which thus suffers fragmentation.

The phenomenon of response of rocks to release of confining pressure, due to unloading, is known as 'dilation'. This is often attributed to the process of exfoliation.

(c) *Growth of crystals* Alongwith water most of the soluble constitutents of the rocks and minerals enter the rocks through its fractures and joints. With the evaporation of water, the salts it contains start crystallising. As the crystals grow, they exert large expansive stresses that result in the weakening of the rocks and ultimately their fragmentation. This process operates extensively in dry climates. During the dry period, due to strong heating by the sun, water deep within the rock is drawn to the surface by capillary force. This water carries dissolved mineral salts. At the surface, the water evaporates and whatever material is in the solution gets crystallised. The growth force of these crystals can cause a disruptive mechanical effect, by which capillary cracks are widened, enlarged resulting in the disintegration of rocks. Crystals of sodium chloride, calcium sulphate (gypsum), magnesium sulphate, sodium carbonate, calcium carbonate, various phosphates, nitrates, alum etc. grow in this manner.

(d) *Freezing of water* Water, as we know, expands by about 9 percent in volume when it freezes. Water trapped in the pores, fissures and crevices of rocks, when freezes, exerts enormous pressure on the walls of the fissures. In case the freezing water is completely confined, it exerts pressures of thousands of pounds per square inch which far exceeds the tensile strength of most of the rocks. Preliminary freezing in the upper part of the water filled crevices may form closed systems, in which further freezing of water produces the pressure capable of breaking the rock that confines the water. Therefore, alternate freezing and thawing is the most effective process of rock-disintegration. This phenomenon is often termed as *Frost-weathering*. This is particularly noticed in the high mountains, where during the day-time the temperature rises high enough to melt some ice and snow and its subsequent freezing in the night when the temperature drops. This is also known as frost-wedging.

(e) *Frost Action* Water present in the ground when freezes, layers of ice tend to form and more water is drawn to the freezing layer of ice by capillary action. Thus the thickness of the ice-layer increases forcing the soil above it upward. In this way, the upheaved soil is disrupted by the expansion of freezing water. This is also known as *Frost heaving*. Three conditions mostly favour this process:

(i) presence of pores in the ground;

(ii) presence of water in the pores; and

(iii) appropriate temperature.

Apart from the various factors already described, it is also noticed that certain rocks suffer volume changes when saturated with moisture; some rocks disintegrate under a state of alternate wetting and drying and that soil colloid may have the power to loosen and lift rock-grains with which they come in contact, thus brings about the physical weathering of rocks.

CHEMICAL WEATHERING

It is a process of mineral alteration, which consists of a number of chemical reactions, whereby the *primary minerals* (i.e. the original silicate minerals of igneous rocks) are converted into new compounds, the *secondary minerals*, which are stable in the surface environment. The sedimentary and metamorphic rocks are also substantially affected by the chemical processes of weathering. Chemical weathering is the result of interaction of rocks of the superficial layers of the lithosphere with chemically active constituents of the atmosphere. Most important of these constituents are water, carbon dioxide and oxygen. The effectiveness of these constituents depends on the following factors:

(i) *Size of the particles* The smaller the particles, the greater is the surface areas which may come in contact with the atmosphere. Since substances react chemically along surfaces, the greater the surface area, the more pervasive is the chemical weathering.

(ii) *Composition of the rock* According to the general order of stability of minerals, as tabulated by Goldich (1938), quartz is very stable. As such, rocks composed primarily of quartz decompose very slowly. Similarly, rocks composed mostly of ferromagnesian minerals like olivine, augite etc. are highly susceptible to chemical weathering.

(iii) Favourable temperature and humidity condition.

The chief chemical weathering processes are :

(a) Hydration;

(b) Hydrolysis;

(c) Oxidation;

(d) Carbonation; and

(e) Solution

Hydration

The term *hydration* refers to the chemical union of water with a mineral. In this process, certain minerals take up water, which lead to a change in the mineral composition of rocks. Due to the absorption of water, the minerals expand, causing more stresses within the rock. The most well-known examples of hydration occurring in nature is the altering of anhydrite to gyspum, hematite to limonite as shown below:

(i) $CaSO_4 + 2H_2O \rightarrow CaSO_4.2H_2O$ (Gypsum)
(anhydrite) (Water)

(ii) $Fe_2O_3 + nH_2O \longrightarrow Fe_2O_3.nH_2O$
(hematite) (Water) (Limonite)

It is to be remembered that when a mineral is hydrated, it is the (OH-) ion that is built into the new crystal lattice.

Since the process of hydration causes swelling of some minerals, this in turn produces a considerable mechanical effect on the enclosing rocks and causes even local dislocations.

Hydrolysis

It is the process of exchange reactions between the bases of the minerals and the hydrogen ions of the electrolytically dissociated part of the water. As we know, water is an active chemical agent because it is always to some extent dissociated into H^+ and $(OH)^-$ ions. The acidity and alkalinity of water is measured by the concentration of hydrogen ions i.e. the pH value. Water having the pH value greater than 7 is alkaline and less than 7 is acidic in nature. The higher the pH value, the more dissociated is the water and the stronger its action as a chemical weathering agent.

The most characteristic example of hydrolysis is that of the potash-feldspar(orthoclase), where the reaction can be shown as

$$\underset{\text{(orthodase)}}{KAlSi_3O_8} + \underset{\text{(Water)}}{HOH} \longrightarrow \underset{\text{(alumino-silicic acid)}}{HAlSi_3O_8} + \underset{\text{(potassium-hydroxide)}}{KOH}$$

But since carbon dioxide (CO_2) is almost invariable present in the atmosphere, the rain water usually contains dissolved CO_2. Water is more readily dissociated when it contains free CO_2. This reacts with the potassium hydroxide giving rise to potassium-carbonate and water.

$$H_2O + Co_2 \longrightarrow \underset{\text{Carbonic acid}}{H^+ + (HCO_3)^-}$$

$$2KOH + H_2CO_3 \longrightarrow K_2CO_3 + 2H_2O$$

The alumino-silicic acid, formed through the hydrolysis of potash-feldspar, is unstable and breaks down with the formation of clay mineral and collidal silica.

Another example of hydrolysis is the formation of magnesium hydroxide from olivine, as

$$\underset{\text{(Olivine)}}{MgFeSiO_4} + \underset{\text{(Water)}}{2HOH} \longrightarrow \underset{\text{(Magnesium -hydroxide)}}{Mg(OH)_2} + \underset{\text{(Silicic acid)}}{H_2SiO_3} + \underset{\text{(Ferrous oxide)}}{FeO}$$

Oxidation

The process of oxidation involves the chemical union of oxygen atoms with atmos of other metallic elements. Thus the minerals are altered with the production of oxides. The free oxygen of the atmosphere and of the air dissolved in water is a most active chemical reagent. It acts actively upon many minerals, especially those containing iron. The ferromagnesian minerals like pyroxenes, hornblende and olivine etc. rapidly undergo oxidation in the surface conditions, producing a brown crust consisting largely of oxides of iron. The oxidation effects are most conspicuous in warm, moist climates.

An example of oxidation of *pyrite* may be shown as follows:

$$\underset{\text{(Pyrite)}}{FeS_2} + \underset{\text{(Oxygen)}}{nO_2} + \underset{\text{(Water)}}{mH_2O} \rightarrow \underset{\text{(Ferrous sulphate)}}{FeSO_4} \rightarrow \underset{\text{(Ferric sulphate)}}{Fe_2(SO_4)_3} \Rightarrow \underset{\text{(limonite)}}{Fe_2O_3.nH_2O}$$

Carbonation

Carbon-dioxide is a gas and is a common constituent of the earth's atmosphere. This is the process by which carbondioxide is added to minerals to form certain carbonates. Rain water in course of its passage through the atmpsphere, dissolves some of the carbon-dioxide present in the air. It thus turns into a weak acid called carbonic acid

$$\underset{\text{(Water)}}{H_2O} + \underset{\text{(Carbon-dioxide)}}{CO_2} \longrightarrow \underset{\text{(Carbonic acid)}}{H_2CO_3}$$

This water containing carbondioxide is capable of reacting with several minerals. It dissolves carbonates with relative ease. Thus, sometimes a limestone may be entirely removed leaving behind the insoluble materials it contains.

$$CaCO_3 + H_2O + CO_2 \longrightarrow Ca\,(HCO_3)_2$$

The process of carbonation is more effective with minerals containing alkali metals like sodium and potassium as well as calcium and magnesium. A characteristic example of decomposition of the mineral feldspar through the process of carbonation, can be illustrated as follows:

$$\underset{\text{(Orthodase)}}{2KAlSi_3O_8} + \underset{\text{(Carbonic acid)}}{H_2CO_3} + \underset{\text{(Water)}}{H_2O} = \underset{\text{(Potassium carbonate)}}{K_2Co_3} + \underset{\text{(Kaoclinite)}}{Al_2Si_2O_5(OH)_4} + \underset{\text{(Silica)}}{4SiO_2}$$

In this illustration, some of the components of feldspar are removed in solution leaving the rest in the form of clay (Kaolinite is one of the common mineral in clay). The effect of this process is well noticed in the limestone or chalk areas in the humid regions of the world.

Solution

Some of the minerals get dissolved by water and thus removed in solution. Even though the process of solution is wide spread in natural environments, pure water is not an effective solvent for any of the common rocks. Solution takes place at different rates for different rocks. Maximum degree of solution is found to be with chlorides (halite (NaCl), sylvite (KCl);) etc. Sulphates and carbonates are less soluble. But the joint action of carbon dioxide and water enhances the effectiveness of the process of solution. As for example, although calcite ($CaCo_3$), the chief mineral in limestone is very slightly soluble in pure water, it has been noticed that vast quantities of limestone have been dissolved and carried away by the water containing carbon dioxide.

It has also been found that silica eventhough is a mineral of high stability under surficial condition undergoes solution in alkaline fluid.

The process of removal of soluble material from the rocks in solution by percolating water, is termed *Leaching*. These processes of chemical decomposition of rocks go on side by side resulting in the weathering of the rock masses.

BIOLOGICAL WEATHERING

The processes of weathering which are mainly related to the activities of plants, animals and organisms like bacteria etc. are known as 'Biological weathering'. It involves the role of plants and animals in the breaking down of rocks through mechanical ways as well as in the decomposition of rocks.

Role of Plants and Animals in the Physical Breaking Down of Rock

(a) Plant-roots, growing between jointed blocks and along minute fractures between mineral grains, exert an expansive force tending to widen the existing openings and sometimes create new fractures.

(b) Insects like earth-worm, snail etc. and burrowing animals like rodents loosen the soil cover and create suitable conditions for the various external agencies to have their own action on the underlying rocks that lead to rock weathering.

Bio-chemical Process of Weathering

(a) The role of organisms in chemcial weathering depends much on their ability to assimilate various elements from the rocks and to evolve various organic acids etc. Plant roots dissolve various constituents of the rocks with the acids evolved by the tips of their roots. Besides, certain groups of bacteria algae and mosses breakup rock-forming silicates directly, removing from them elements like silica, potassium phosphorous, calcium, magnesium etc. that they need as nutrients. This facilitates rock-weathering.

(b) Bird droppings have been stated to be capable of weathering limestones.

(c) With the decay and degeneration of the dead remains of plants and animals, chemically active substances are produced like carbon dioxide, humic acid, together with traces of ammonia, nitric acid etc. These substances are capable of bringing about rock-weathering.

(d) Man himself is an important agent of destruction.

All the processes of weathering, physical, chemical and biological, are interconnected and act simultaneously.

WEATHERING PRODUCTS AND GEOLOGIC FEATURES

The products of weathering usually include the following

(i) The first products of weathering is a mantle of broken and decomposed material of varying thickness and composition, called the *regolith* which covers the areas except those from which it is removed as soon as formed.

(ii) Soluble salts, which are produced, are carried away along with the transporting media in solutions.

(iii) Colloidal substances like $Al(OH)_3$ and $Fe(OH)_3$ which are the products of weathering, are carried away by ground water and streams.

(iv) Insoluble products, which include clay minerals, quartz grains, undecomposed feldspars and some resistant minerals like zircon, tourmaline, quartz etc. are usually found at the site of weathering and later transported to the sites of deposition.

All the weathering products are grouped into two categories as:

(a) Transported or mobile and

(b) Residual or sedentary

The products of weathering belonging to the first category are those which are transported over varying distances by mechanical, chemical and biochemical means.

They include various types of clastic and nonclastic sedimentary rocks. The products belonging to the second category are those which are accumulated at the site of destruction of rocks. These are usually the insouble products of rock-weathering and they still mantle the rocks from which they have been derived.

The most important and immediate results of the weathering processes is the formation of soil. This may be residual or transported.

Weathering also gives rise to residual concentration of minerals of economic importance. Residual concentration results in the accumulation of valuable minerals when the undesired constituents of rocks and minerals are removed during weathering. Bauxite deposits, iron deposits (in the form laterite concretion), manganese, nickel and cobalt deposits are formed in this fashion.

Weathering gives rise to infiltration-deposits through oxidation and secondary enrichment. In such processes, oxidation, solution (leaching) and consequent downward movement of the valuable ore minerals lead to their concentration below the water table due to the existence of reducing condition there. Copper deposits are sometimes formed in this manner.

GEOLOGICAL FEATURES

Differential weathering gives rise to features like honeycombed rocks, hollows and niches in rock walls, weathering pits on lime-stones, spheroidal boulders (formed due to exfoliatin or spheroidal weathering), tors (a tor is a residual mass of rock usually capping a hill, slope or spurs). Weathering aided by mass-wasting produces landforms as *Talus slopes* or *scree.*

In conclusion it may be stated that weathering of rocks is not an isolated process. It is closely related to the activities of the other exogenous agents which bring about the degradation of the earth's surface. Weathering lowers the surface of the rock being affected and produces broken or disintegrated rock material which can be transported more easily by the geologic agents like running water, ice or wind etc.

19

SOIL AND ITS CONSERVATION

Next to water and air, soil is most essential to our very existence on the earth. Soils are the basis of support for most life, and a source of nutrients for marine life and fresh water. As a natural resource, soil is of immense value to man. In the agricultural orient the distribution and density of population have conformed to the persisting patterns of soil fertility and productivity.

Soil is basically broken-down rock materials and consists of decomposed rock debris and decayed organic matter(humus) which have been produced by weathering. According to Arthur Holmes *From a geological point of view soil may be defined as the surface layer of the mantle of rock-waste in which the physical and chemical processes of weathering co-operate in intimate association with biological processes* (Principles of Physical Geology, Second Edition Completely Revised, the English Language Book Society & Nelson). Arthur N. Strahler and Alan H. Strahler defines soil as a *natural surface layer containing living matter and supporting or capable of supporting plants.* (Modern Physical Geography, John Wiley & Sons, Inc., Newyork). Chester R. Longwell & Richard F. Flint defines soil as—*Soil is that part of the regolith that will support rooted plants.* (Introduction to Physical Geology, 2nd Edition, John Wiley & Sons, Inc. New York, London). From the above definitions, it may be stated that—

Soil is the surface layer of the earth restricted to land, consisting of a layer of broken-down, fine and loose rock material, produced by the weathering processes, mixed with decayed vegetation and other organic matter. Soil may be found on top of the parent rock or at some distance from it after transport and supports vegetation.

The science which deals with the study of the soil is known as *Pedology* and the process of soil formation is called pedogenesis.

The Soil Profile

The pedologists have distinguished a number of layers or horizons within the soil. A vertical section made through a soil reveals a series of more or less distinct layers. These layers from the surface down to the unchanged parent material is called the *soil-profile*, which is characteristic of the residual soil (i.e. the soil developed on top of the parent rock), where there is a gradual transition from the top soils to the partially decayed rock and finally to the unaltered rock). But a transported soil does not show any such characteristic.

A simple soil-profile shows three distinct layers designated as A, B and C- layer. The upper layer containing most of the organic material is called the A-layer or horizon, which is commonly known as the *top-soil*. This is the horizon of maximum biological activity.

The layer below the A-horizon is the B-horizon, which is poor in organic content and rich in clay. This layer is regarded commonly as 'sub soil' Mineral matter removed from the A-horizon through solution are precipitated in the B-horizon. High concentration of clay minerals in the B-horizon may be due to mechanical removal of colloidal clays in suspension by the descending soil water. Deposition of the ferro-humus material along with the silt and clay particles sometimes form a layer which is dense, tough and well cemented and is called a *hard pan* or *clay pan*.

The C-horizon lies below the B-horizon.

This horizon contains remnants of the parent material and is little affected by biologic activity. However, it is affected by physical and chemical processes. This horizon grades downward into the unaltered parent rock. Bedrock underlying the C-horizon is designated as *R-horizon.*

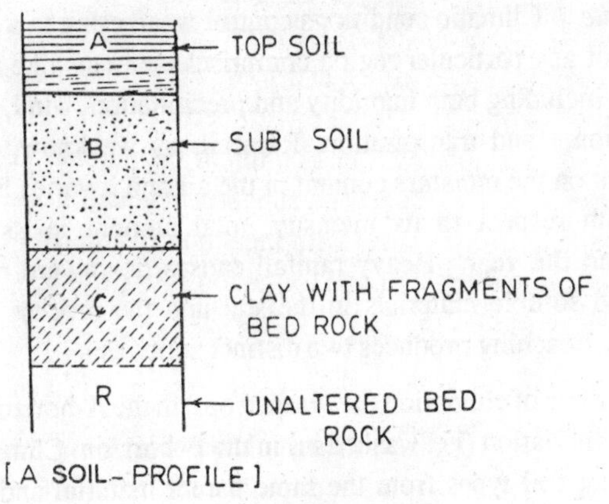

[A SOIL- PROFILE]

FIG.46.

SOIL FORMATION

The natural processes of soil formation are very slow; and are due to a combination of several factors such as 1. Parent rock material, 2. Climate, 3. Plant and animal life, 4. Local topography 5. Time etc. Most of these factors are interdependent. The processes of soil formation are most intimately associated with the weathering processes and the factors indicated above also determine the characteristics of the soil.

1. *Parent rock material* These are the bedrocks on which the soil develops. The processes of mechanical disintegration and chemical decomposition which constitute weathering proceed side by side resulting in the breaking down of the bedrock into a mixture of soluble and insoluble materials. While the soluble matter is removed in water, the insoluble residue forms a framework for the development of soil. Since the parent material or the bedrock is composed entirely of materials, it provides the inorganic constituents to the soil. The mineralogical composition of the bed rock from which the soil is derived, determines the rate of physical and chemical weathering as well as the composition of the soil. As we know, rocks composed of resistant minerals are slowly affected by weathering processes. Texture and structure of the parent

material also determine the rate of weathring and the degree of retention of plant nutrients in the soil.

2. *Climate* Climatic conditions control weathering to a great extent. Climate of any particular region comprises elements like temperature, moisture including both humidity and precipitation, wind, air pressure, evaporation, and transpiration. Rainfall, as we know, is primarily dependent on the moisture content in the air and it varies from place to place with respect to its intensity, total amount in its distribution throughout the year. Heavy rainfall causes downward movement of water and soluble materials in the soil and the process is known as leaching. Leaching produces two distinct zones as:—

(i) Zone of eluviation (i.e. washed out) in the A-horizon and (ii) Zone of illuviation (i.e. washed in) in the B-horizon. Climate gives rise to different soil types from the same parent material and also widely different parent materials may produce similar soils in one climatic context.

Apart from rainfall, temperature plays a significant role in the weathering process. Temperature and moisture not only affect the rate of chemical weathering but also of bacterial activity, on the parent rock material in the process of soil-formation.

3. *Plant and Animal life* Many pedologists believe that soil is a biological phenomenon and that the plants play the leading role in the process of soil formation.

Plants promote disintegration of rocks by the growth of trees in cracks or joints, thus wedging off large and small fragments of rocks. Roots often penetrate into the crevices of rocks exert on expanding force on the side walls.

The biochemical activity of plants includes the extraction of various mineral substances, water and the necessary elements of nutrition on the one hand, and on their death and decomposition they contribute towards the accumulation of organic matter in the soil, on the other. Dead plants contribute to the humus content of the soil and the process of humification releases carbon dioxide and organic acids together with traces of ammonia and nitric acid etc. which often speed up the decomposition of the mineral matter and accelerates soil-formation.

The microflora such as bacteria, algae and fungi contributes significantly to soil formation. But the activities of both bacteria and fungi are related to climatic conditions. In cold climates bacterial activity is limited whereas it is very intense and rapid in warm, moist climates. Bacteria are also involved in the nitrogen and sulphur cycles. Even the bacteria sometimes cause the quicker decay of neighbouring rock surfaces.

Burrowing animals are effective soil makers. They make the soil and softer rocks porous and spongy and thus make them more readily susceptible to weathering and erosion. It has been estimated that earthworms completely work over a soil layer of 6 to 12 iches thick every 50 years. They extract vegetable matter from the soil by eating their way through it. As the soil passes through their bodies, it is subjected to mechanical and chemical modification.

4. *Local topography* It affects the character of the soil profile. True soils with a full profile can develop only on fairly flat surfaces where erosion is slow; whereas on steep slopes the profile never becomes completely developed as erosion removes the products of weathering as soon as they form. On flat upland surfaces, a thick soil is formed, often with a layer of clay, but it is well leached as uplands also attract heavier rainfall. On flat-bottom lands in the flood plains, there is dark coloured, thick soils, since the flat bottom lands are poorly drained.

5. *Time* The development of a matured soil profile requires time. Soils are less well-developed if the soil-forming processes have not been in operation for an adequate time period for a fully developed soil-profile, in most places, it needs several thousands years.

TYPES OF SOIL

A number of classification have been made on different basis by the agricultural scientists, geographers and geologists. The objects of classifying the soils differ in different classifications. The geographic and geologic classification of soil are only important in the present context of our study.

Geographic Classification

Geographers have classified soil in terms of their areal distribution over the earth's land surfaces which are linked with the climates, parent materials etc. As we know, under similar climatic conditions similar soil types develop.

The U.S. Department of agriculture proposed classification system in 1938, which is much simpler and recognises the existence of three orders of soils

1. Zonal Soils
2. Intrazonal Soils
3. Azonal soils

Zonal soils are by far the most important and widespread soils. These are exhibiting a mature and well-developed soil profile indicating the fullest play of various soil-forming factors. They develop on well-drained areas, on parent material which has ramained in the original place for a sufficiently long time to have been affected by various soil-forming processes.

Intrazonal soils are developed under conditions of poor drainage, on regolith where soluble salt contents are high. Soils of bog areas and alkali flats are examples of intrazonal soil. All of them have distinct profile characteristics.

Azonal soil lack well-developed soil profiles which may be due to non-availability of sufficient time for them to develop fully or due to the location on very steep slopes which prohibits profile development. Alluvial soils, dune sands, lithosols (i.e. mountain soils on steep slopes) and organic soils, which develop on peat bogs are examples of azonal soils.

The Major Soil Groups of the World

The zonal soils are subdivided into the following types on the basis of climatic zones.

(a) *Soils of the humid tropics* The humid tropical regions of the world are found close to the equator, where the average temperature is 25°C with little seasonal variation and a rainfall of over 2000 mm each year. These climatic conditions favour the formation of laterites, which are the common tropical soils. These are the leached, hard, concentrated horizons of iron and aluminium oxides and are used in making bricks.

(b) *Soils of humid temperate regions* In middle latitudes, leaching is the dominant soil-forming process. Two main groups of soils occur in this climatic zones, as Podzols and Brown earths.

Podzols are the dominant soils of the zone between 50° North and the Arctic circle. They develop under conditions of a cold winter and an adequate precipitation spread through out the year. They are found mainly in the northern coniferous forest belt and on infertile, sandy and gravelly areas in warmer climates. In this zone precipitation ranges between 500 to 1000 mm and evaporation is low, which encourages leaching. The leaching process carries iron and aluminium compounds out of the surface horizons and makes them rich in silica.

Brown earth's are mostly found in the deciduous forests, where the winter is shorter and rainfall is more evenly distributed throughout the year. Here leaching is less intense and the soils are more fertile than podzols.

(c) *Soils of seasonally wet regions* These regions include three groups

(i) Areas with wet winters i.e. Mediterranean type regions which are characterised by cyclonic winter rain causing leaching. Cinnamon soils are the examples.

(ii) Tropical areas with wet summer, where there is maximum rainfall in the summer. Ferruginous soils are well-developed in such regions.

(iii) Temperate areas with wet summer which occur in the semi-arid middle-latitude steppe lands of North America and Asia. These regions have a low rainfall(mostly in summer) and cold winters. Chestnut brown soils are the most characteristic soils of this climatic region. *Chernozems* are the important soils developed in the south-central parts of USSR. Such soils are extremely fertile.

(d) *Soils of arid areas* Here the rainfall is very less and the area occurs between 30° and 50° North latitudes in the northern-hemisphere. In this region day time temperatures are quite high and humidity is low and there is almost no leaching. There is little profile development in the soil e.g. 'grey desert soils'.

(e) *Soils of cold areas* These are the soils developed in a tundra climate where the soils are frozen for a large part of the year. Here the drainage conditions are poor and boggy. These are also known as Tundra soils.

The intrazonal and azonal soil types are not of much significance.

Geologic Classification

On the basis of the mode of occurrences and the natural agents involved, soils are commonly classified into two categories as

1. Sedentary soils
2. Transported soils

1. *Sedentary soils* These are also known as the residual soil. They occur directly on top of the parent rock. These are the residues left as insitu after weathering followed by transportation and consist of the insoluble products of rock weathering, which have escaped distribution through transporting agencies and which still mantle the rock from which they have been derived. As such, the chemical composition of residual soils is defined principally by the nature of the parent rocks. The residual soils are often rich in humus. Laterites form the best example of residual soil.

Under marshy condition, accumulation of organic matter like peat gives rise to an organic type of soil. This is known as 'cumulose soil' and is included in the category of sedentary soils.

2. *Transported soils* These are also known as drifted soils and this category of soil includes all those soils that have been deposited at places far from the parent rocks after being transported by the geologic agents. On the basis of the transporting agencies involved, these soils are classified as follows

(a) *Colluvial soils* Under the influence of gravity material are removed from the mountains and get accumulated at the base of the steep slopes. The soils thus formed are stony and are never stratified.

(b) *Alluvial Soils* These soils are generally confined to river basins and coastal plains and are mainly deposited by rivers. These are very fertile and supports vegetation.

(c) *Glacial Soils* These soils are transported and deposited by glaciers. Boulder clay or till forms good soil at times.

(d) *Aeolian soils* The wind-borne sediments, composed chiefly of silt and clay-fractions form scanty but fertile soil at times. The loess deposits form good examples of aeolian soils.

(e) *Lacustrine soils* Materials transported by rivers and glaciers, collected in the lake basins, form soil in due course when the lakes dry up. They are stratified and are rich in organic matter.

Apart from the above important types of soil, there are sandy, coarse-grained soils formed from the sediments deposited in the coastal regions or on continental-shelf area, which constitute the *marine soils.* Similarly, soil derived from the pyroclastic materials are known as *Volcanic soils.*

SOILS OF INDIA

The vast size of India, along with the diversity in the natural environment of its various parts gives rise to a vast variety in the soil cover. As we have already described, the formation of different types of soils are mainly dependent on the nature of parent material, climate, general topography of the particular region and activity of organisms. These factors vary in their degree of operation from one place to the other and thus result a number of soil types in India. While the soils in the Extra-penisular and Indo-Gangentic plains are quite young, those in the Peninsular parts are old and highly matured.

The major soil types in India have been recognised as follows:—

— Alluvial soils, which are well developed in the Indo-Gangetic plains as well as in the coastal deltaic regions

— Regur or black cotton soil, which are concentrated over the Deccan Lava tract that includes parts of Maharashtra, Madhya

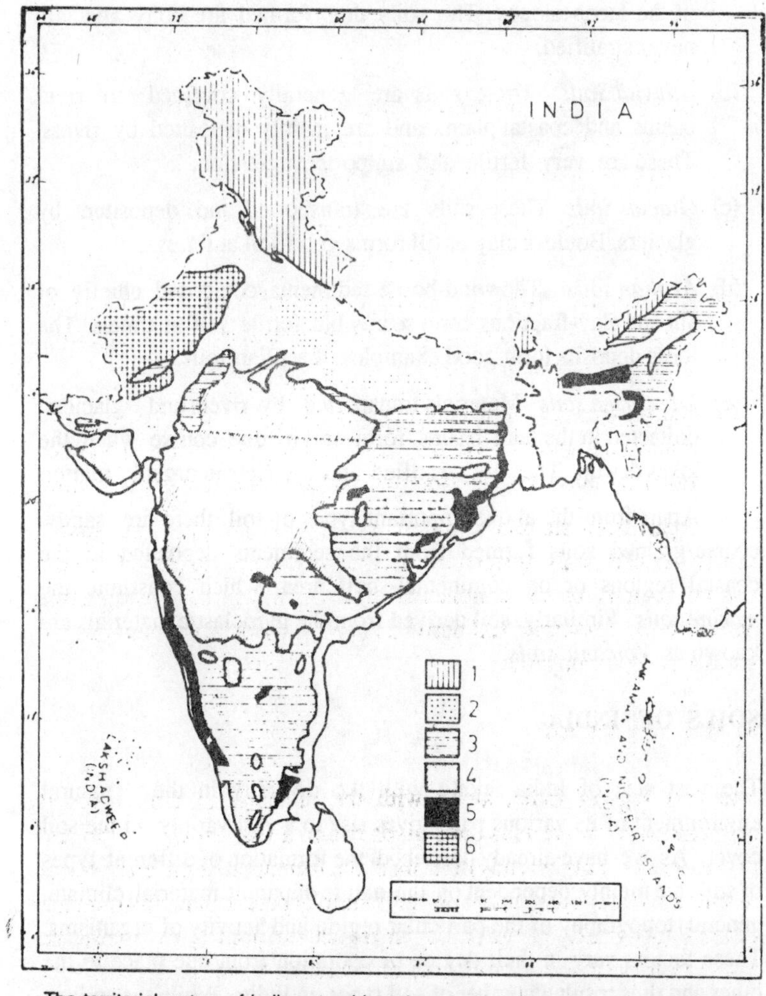

The territorial waters of India extend into the sea to a distance of twelve nautical miles measured from the appropriate base line

India -- Major Soil Types

1. Mountainous Soils 2. Alluvial Soils
3. Red Soils 4. Black Soils
5. Laterite Soils 6. Desert Soils

FIG. 47. SOILS OF INDIA

— Pradesh, Gujurat, Andhra Pradesh and some parts of Tamil Nadu. It is highly fertile.

— Red soils are due to a wide diffusion of iron in ancient crystalline and metamorphic rocks. Red soils cover almost the whole of Tamil Nadu, Karnataka, parts of Andhra Pradesh, Orissa and eastern parts of Madhya Pradesh and south-eastern part of Maharashtra.

— Lateritic soils are commonly found in Karnataka, Kerala, Tamil Nadu, Maharashtra, Madhya Pradesh and the hilly areas of Orissa and Assam.

Desert soils are found in Rajsthan.

Apart from the above types, the soils in the mountainous regions of the country also form a significant kind of soil of India.

SOIL EROSION AND ITS CONSERVATION

Soil is one of the most important natural resources of man. Soils are essential for man for growing crops, fodder and timber. Once the fertile portion of the earth's surface is lost, it is very difficult to replace it. In India, the destruction of the top-soil has already reached an alarming proportion. Land degradation problems have resulted in increasing depletion of the productivity of the basic land stock through nutrient deficiencies. In addition to the direct loss of crop producing capacity, soil erosion increases the destructiveness of floods and decreases the storage capacity of water in reservoirs. It is therefore essential that the soils should not be allowed to wash or blow-away more rapidly than they can be regenerated, their fertility should not be exhausted and their physical structure should remain suited to continued production of desired plant materials.

Protection of land from further degradation, adoption of various conservation measures, including reclamation and scientific management of available land stock is very important for a country like India to achieve higher productivity of food, fodder, fuel and industrial raw materials on a substantial basis. Besides, demand for land for providing social priorities such as shelter, roads, industrial activities is increasing at a very fast rate with the rise in population and very often good agricultural and forest lands are being diverted to such use. It is, therefore, necessary to keep soil in place and in a state favourable to its highest productive capacity.

Soil Erosion

The process of destruction of soil and the removal of the destroyed soil material constitute soil erosion. According to Dr. Bennett "the vastly accelerated process of soil removal brought about by the human interference, with the normal disequilibrium between soil building and soil removal is designated as soil erosion".

Types of Soil-Erosion

Erosion of soil by water is quite significant and takes place chiefly in two ways (a) Sheet erosion, (b) Gully erosion.

(a) Sheet movement of water causes sheet erosion and depends on the velocity and quantity of pronounced surface runoff and the erodability of the soil itself. In such cases, the soil is eroded as layers from the hill slopes, sometimes slowly and insidiously and sometimes more rapidly. Sheet erosion is more or less universal on:—

- all bare follow land,
- all uncultivated land whose plant cover has been thinned out by over grazing, fire or other misuse, and
- all sloping cultivated fields and on sloping forest, scrub jungles where natural porosity of soil has been removed by heavy grazing, felling of trees or burning etc.

The particles loosened and shifted by the rain drops are carried down slope by a very thin sheet of water which moves along the surface. The impacts of the raindrops increases the turbulance and transporting capacity of this unchannelized sheetwash which results in the uniform skimming of the top soil. Sheet erosion is considered as dangerous as it may continue for years but may or may not leave any trace of the damage. Sheet erosion is common in the Himalayan foothills, in Assam, Western ghats and Eastern ghats.

When sheet erosion continues unchecked, the silt laden run-off forms well-defined minute finger shaped grooves over the entire field. Such thin channelling is known as 'rill-erosion', which is active over wide areas in Bihar, Uttar Pradesh, Madhya Pradesh and in semiarid areas of Maharashtra, Karnataka, Andhra Pradesh and Tamil Nadu.

(b) *Gully erosion* On a gentle slope, adequately covered by vegetation, clay soil will resist erosion to a great extent and the water forms small rivulets which can then erode deeper. The rivulets in turn join together to form larger channels until gullies are formed gradually deep gullies cut into the soil and then spread and grow until all the soil is removal from the sloping ground This phenomenon once started and if not checked, goes on extending and ultimately the whole land is converted into a bad-land topography. Gully erosion is more common in areas where the river system has cut down into elevated plateaus so that feeders and branches carve out an intricate pattern of gullies. Apart from this, it also takes place in relatively level country whenever large blocks of cultivation give rise to concentration of field run-off.

Wind Erosion

It occurs in dry climatic areas having a sparse and low vegetation cover on mechanically weathered, loosened surficial material. Dust storms are the principal agents of wind erosion. The top soil is often blown off from the surface rendering it infertile. Besides, with the decrease in the wind velocity coarse sand particles get deposited in some areas covering the existing soil and rendering it unproductive.

Causes of Soil-Erosion

While the nature takes from 100 to 400 years to build one centimetre of top soil, man can and often does destroy it almost overnight by haphazard land use and improvident husbandry. Irrational methods of cultivation, deforestation, destruction of natural vegetation due to overgrazing by pasturing animals etc., accelerate denudation. Besides, failure of rains, floods, depopulation and loss of cattle caused by famine and pestilence, disturbance caused by war and interference with or change in the natural drainage system have had their deleterious effect on the soil at some time or the other.

(a) *Irrational Methods of Cultivation*

(i) *Faulty method of cultivation* Particularly on the steeper slopes when the virgin land is ploughed and naked soil is exposed to the rain, the loss of fertile soil is enormous. The potato cultivation in the Himalayas and the Nilgiris, where the rows run straight up and down hill, causes an abnormal rapid loss of soil.

(ii) *Shifting cultivation* It is a primitive form of soil utilisation. In shifting cultivation, framers grow food only for themselves and their families. In this system of farming a patch of forest is selected. Its tree and bushes are than cut and burnt down on the ground in order to clear room for a field. The ground is then lightly ploughed and seed is sown broadcast and racked into the soil at the first fall of rain. The soil gives rise to a better yield as it is immensely fertile owing to the wood ashes and accumulated humus. After two or three years' crop, when the fertility of the soil is seriously reduced, the people again change their land of cultivation. Thus the essential feature of shifting cultivation is the rotation of fields rather than crops. As a result more and more land are exposed to erosion.

(iii) *Nature of crop grown* In India, as it has been noticed the dry crop producing regions (such as millet, maize, potato, tobacco, cotton and even wheat growing region) of high temperature, low humidity and scanty rainfall are attended by heavy loss in soil specially at the time of a heavy shower. In contrast to these, rice, jute and sugar cane account for very little loss of soil.

(b) *Deforestation* The cover of the vegetation not only reduces the velocity of surface runoff, but also binds the soil particles through the roots and increases its strength. Thus vegetation cover protects the soil from the attack of erosional processes. Deforestation leads to increased runoff of rain water and its diminished seepage and storage in the soil. The structure of the soil suffers, the runoff increases which loosens the soil and carry it. Once the channel is filled up with the load, it causes devastating floods at times.

In a natural forest, the force of rain is checked by the leaves of trees and thick carpet of vegetation, while surface covering of soil and the humus soak up the rain water like sponge and let it sink into the ground to emerge afterwards as springs and streams. When the rain falls gently, the whole is absorbed and violent floods in the stream is lessened but deforestation makes the problem more severe.

(c) *Over-grazing* Over-grazing by pasturing animals like cattle, goats and sheep makes the grass cover on the soil worn and thin. As a result, rain drops begin to fall directly on the soil and clogging up the pores with mud and form an impervious crust which reduces infiltration into the soil and increases surface runoff. This leads to an increase in the area of bare ground. As a consequence, they fall easy victims to various erosional processes.

Apart from the above causes, the systems of farming, size of the farm tenancy, tenant-landlord relationship etc. also constitute the socio-economic factors of soil erosion.

Factors affecting Soil-Erosion

There are a number of physical factors responsible for the depletion and erosion of soil which may be described as follows:—

(i) *The concentration of rainfall* The more concentrated and intensive the showers, the more forcefully they strike the surface and greater the runoff erosion.

(ii) *Relief of the ground* On a steeper slope, soil is washed away much more rapidly than on the gentler slope.

(iii) *Nature of the soil* The structure, texture and organic matter contents of the soil also affect soil erosion. Light open soils lose more silt than heavier loams. Heavy black cotton soils, which swell up when wet are probably not denuded as the lighter soils found in the peninsula. The soft shale and sandstone erode more rapidly than limestones, granites etc.

(iv) *Carrying capacity of running water* The eroding capacity as well as the carrying capacity of the water is closely associated with its velocity.

(v) *Nature of vegetation cover* Thick vegetation cover retards the soil erosion whereas the greatest loss due to soil erosion occurs when the ground is bare of vegetation.

Effects of Soil Erosion

(i) Heavy floods in the rivers.
(ii) Lowering of sub-soil water level.
(iii) Water logging and decrease in the crop yields.
(iv) Destruction of tender vegetation
(v) Failure of dams due to silting.

SOIL CONSERVATION

Various methods have been practised to check soil erosion and to improve soil fertility. Methods by which soil is prevented from being eroded constitute what is known as soil conservation. In order to reverse the process of land degradation and resource depletion, appropriate soil conservation measures are to be adopted for bringing more areas under crops, forests and grasses. All methods of soil conservation aim at reducing the amount and velocity of surface runoff and of the erodability of the soil. Some of the important measures usually adopted are as follows:—

1. *Agronomic methods* These method are employed in protecting the top-soil by special method and scheme of crop cultivatiion. They include—

(a) Rotation of crops.
(b) Contour farming.
(c) Strip farming.

2. *Mechanical methods* They include contour terracing, under which a series of properly spaced ridges and drinage channels are formed along contours by construction of suitable mounds of earth.

Apart from this, various types of artificial channels (ditches) are exacavated at suitable locations for the removal of excess water from approaching the field. Also, dams are constructed for checking the velocity of water.

3. *Change in land use pattern* It consists of growing suitable crops in place of low yielding crop and in making best utilisation of the land by putting it to the use for which it is best fitted.

4. *Common methods* (a) Since vegetation cover protects the soil from erosional processes more effectively, these methods mainly aim at creation of protective surface through afforestation and reforestation, Both these methods are quite effective more specifically on slopes where trees retard the surface runoff and bind the soil.

(b) *Cultivation of soil binding crop* Crop like Peas, Jowar, Bajra, gram and other legumes and fodder plants like guar, alfafa, clover, berseem should be encouraged while erosion inducing and soil depleting crops such as tobacco, cotton, potato, maize etc. should not be cultivated on the eroded land.

(c) Control of shifting cultivation through the following ways—

(i) Reclaiming land and providing irrigation and other inputs and services so as to encourage settled cultivation.

(ii) Identifying areas suitable for plantation crops.

(iii) Development of grass lands and areas for supporting animal husbandry programmes, and agricultural operations as well as commercial forest plantation.

(d) *Control on the grazing of animals* Over-grazing causes a picking of the soil. Therefore preventive measures should be adopted to check the intensive grazing of an area, particularly on the slopes. Fencing of worst affected areas may also be tried.

(e) *Flood control measures* Soil erosion, as we know, is particularly severe where heavy rainfall and steep sloping ground favour the rapid loss of any soil exposed by agriculture. Construction of embankments etc. also protect the land from being eroded.

(f) *Reclammation of ravine lands* Small channels are made on the terraces for safely carrying the runoff water downwards into big ravines. It is accomplished by placing a series of terraces across the slope, one below the other.

To prevent wind erosion, shelter belts of trees are to be planted to check the force of the wind. Ploughing of land at right angles to the direction of wind further serves to prevent wind erosion.

All the above methods are useful in the process of soil conservation.

20

MASS—WASTING

Everywhere on the earth's surface, materials are always acted upon by the force of gravitational attraction. This force of gravity plays an important role in the removal of broken rock debris from high elevations and on steep slopes. Evidence of the down-slope movement of rock and regolith is found almost universally. Abundant evidences show that on most slopes at least a small amount of down-hill movement is going on constantly. The movement of rock debris and regolith on slopes is subject to a large variety of processes, some of which act in a slow and continuous manner whereas others a sudden and catastrophic way.

The various kinds of down-hill movements of coherent masses of rock-debris, occurring under the pull of gravity, are collectively known as 'Mass-wasting' or 'Mass-movement'. It constitutes an important process in denudation of the continental surfaces.

TYPES OF MASS-MOVEMENT

C.F.S. Sharpe made one of the first attempts at classification of the various types of mass-wasting ('Landslides & Related Phenomena,

Columbia University Press, 1938). He recognised four major classes of mass-wasting as follows:

I. Slow-flowage
(a) Soil creep.
(b) Talus creep.
(c) Rock creep
(d) Rock-glacier creep.
(e) Solifluction.

II. Rapid-flowage
(a) Earth-flow
(b) Mud-flow
(c) Dabris avalanche.

III. Landslides
(a) Slump.
(b) Debris slide.
(c) Debris fall.
(d) Rockslide.
(e) Rockfall.

IV. Subsidence.

From the above classification, fundamentally mass-movements occur through 'flowage' and 'sliding', which may be so slow that the movement is imperceptible or too rapid like catastrophic slumping and rockfalls.

SLOW-FLOWAGE

Soil Creep

It is the most widespread, slow, down-hill movement of regolith and soil under the influence of gravity. It usually occurs on gentle-to-medium gradient slopes where the weathered mantle is deep and is particularly common when it is also fine-grained. Here, the down-hill movement of regolith is impreceptibly slow.

A number of processes, each capable of producing only very slight movements combine to cause soil creep. Heating and cooling of the soil, growth of frost needles, alternate drying and wetting of the soil, trampling and burrowing by animals and vibrations of earthquakes all produce some disturbance of the soil and regolith. As gravity always exert a downward pull on every such rearrangement, the particles are urged slowly downslope. Soil creep may operate in unconsolidated materials on slopes less than approximately 35°.

The effects of soil-creep are evident in the consistent leaning of old fences, poles and gravestones.

Among the other types of slow-flowage, there are talus creep, rock creep and rock-glacier creep which involve the downhill movement of the materials of larger sizes, but the process of creeping remains the same in all the cases. While talus-creep is the down-hill movement of pieces of bed-rock ranging in size from tiny chips to sizeable blocks, which rest at a particular angle of repose; rock-creep is a movement of jointed blocks, partly due to soil creep and partly as a result of sliding and rock-glacier creep is a movement of streams of boulders with little soil and only interstitial ice.

Solifluction

The slow, down-hill flowage of water saturated regolith is known as *solifluction*. It occurs in cold lands where there is an annual cycle of freeze-thaw i.e. in frigid climates. During the winter the ground is frozen and partially thaws out in the summer months. When thawing occurs it affects only upper part of the regolith which have been forced up and deranged by frost-heaving while the deeper part stays solidly frozen. Surplus water then cannot drain downward and the thawed layer gets fully saturated with water. This water-saturated regolith layer flows almost imperceptibly down hill. The moving soil and debris takes forms like sheet, lobate or tongue as it moves.

RAPID-FLOWAGE

Rapid flowage is related largely to increasing supplies of water on steeper slopes.

Earthflow

In regions of humid climate,,a mass of water-sturated soil, regolith or weak clay or shale layers may move down a steep slope during a period of a few hours in the form of an earthflow. This is also known as soilflow. The water-saturated material flows sluggishly to form a bulging toe. Some high mountain areas in which bed rock is very weak and easily weathered are afflicted with large earthflow. The uper part of the earthflow undergoes a subsiding motion with backward rotation of the down-sinking mass, whereas the interior and lower parts of the mass move by slow flowage and a toe or lobe of extruded flowage material may be formed. The shortness of the toes which develop downslope are due to the loss of water by drainage and infiltration which eliminates any elements of liquidity in the flow mechanism and tends to extinguish rapid movements. The flowing mass leaves a steep scarp in the rear.

In general, earthflowage occurs only on fairly steep slopes; but large-scale flowage of saturated clay may take place on slopes of less than 1°.

Mudflow

Mudflows are fluid mass of regolith moving in surges down stream channels. They are characteristic of steep, scantily vegetated slopes on which heavy rainfall initiates movement in a thick layer of weathered material. They are mostly found in drier regions.

They differ from earthflows in that (1) they move more rapidly and have a high water content in comparison to the earthflows; (2) they are confined to channels.

They are also known as *mud streams*. They occur where sufficient water is concentrated in the regolith at the head of a valley to overcome the internal cohesion and they move at rates as rapid as a river flow. The weight of water is more important because flow starts when the weight component parallel to the slope exceeds the internal cohesion of the mud and the friction at its base. Many mudflows start after heavy rains in mountain valleys. They have a long, narrow track and spread out on reaching lower ground. Some mudflows are so viscous that they came to a halt in the stream channel and sometimes may block it until the pressure behind it becomes great enough to breach the temporary block.

Debris Avalanche

This is the most rapid form of rapid flowage. In humid regions, soil is developed on steep slopes which are commonly covered with vegetation and tend to be stable. Mass movement here occurs due to some natural disturbance and sliding as well as flow may be involved. Sometimes the materials are so massive and the movement so swift that they are reported as avalanches. They are much like snow avalanches except that rock debris instead of snow makes up the bulk of their mass. Besides, water is an important factor in debris-avalanches.

LANDSLIDES

The term landslide is used widely in a general sense to mean any downslope movement of a mass of regolith or bedrock under the influence of gravity. Here the downslope movement of the mass is always perceptible and is said to involve mainly dry material.

Usually they occur on steep slopes, where hard and heavy rocks overlie softer, or more easily lubricated materials, such as clay or gypsum, in which the bedding or joint-planes dip downward towards the valley.

Slump

It involves the movement of the bedrock mass down a curved slip surface. As the rock-mass moves down it also rotates on a horizontal axis, and therefore the upper surface of the block becomes tilted toward the cliff that remains. In such cases, the mass of rock fails along two surfaces-one is the common surface of shear failure, usually curved outline and the other is the surface separating one block from the other in the immediate neighbourhood. Usually slumping takes place as several small independent units which gives rise to a number of step-like features.

(SLUMP)

FIG.47.

Debris Slide

This is a rapid downslope movement of a mass consisting of regolith of varied texture. They mostly occur on steep slopes covered with scanty vegetation. The amount of water is usually small, thus it is distinguished from debris avalanches. It involves a sliding or rolling motion but not the type of backward rotation as in the case of slumping.

In case of 'Debris falls' material falls from a vertical or overhanging cliff.

Rock Slide

It consists of a rock-mass slipping down the slopes which are usually the planes of weakness like bedding, joint or fault planes.

Rockfall is the free falling or rolling of single masses of rock from a steep cliff and thus involves disintegration of the rock-mass.

SUBSIDENCE

It is a downward movement of the rockmass without any horizontal motion. It takes place mostly due to slow removal of the materials beneath the surface due to chemical action of ground water or when weak plastic materials are overlain by heavy rockmasses. Examples of subsidence may be seen in the collapsed areas of surface due to the underground enlargement of caverns in limestone regions and also sometimes in the mining areas where there is a surface downwarping.

The following factors attribute, to the process of mass wasting:

(a) structural characteristics of a region.

(b) composition of the rocks in a particular region.

(c) climate and vegetation cover.

Mass-wasting like weathering also contributes to the gradation of the earth's crust.

21

GEOLOGICAL WORK OF RIVERS

Water, as we know, is an important agent in bringing about changes near the surface of the earth. This is possible because it is recycled through the oceans, atmosphere and land. Solar radiation causes evaporation of water from the surface-of-water bodies and due to transpiration of plants etc. the moisture content of the atmosphere increases and under favourable conditions this water falls to the earth in the form of rain, hail, sleet or snow. Falling on the earth's surface, atmospheric precipitation is distributed in a number of ways. A part of the water resulting from rain and melted snow enters the soil by infiltration and contributes to the accumulation of under-ground waters; a part is returned to the atmosphere through evaporation; and another part flows over the ground surface as runoff to lower levels due to the pull of gravity.

Runoff, which flows down the slopes of the land, may be represented in the equation form as:—

Runoff in stream = Precipitation-Loss (i.e. Infiltration + Evaporation).

Runoff is of two types viz. Overland (or sheet) flow and Channel (or stream) flow. In the overland flow, the rain and melt waters are distributed over the surface more or less evenly and this occurs only on smooth slopes. They can wash down only the finer weathering products from the surface. Whereas, in the case of channel flow, the water occupies a narrow channel confined by lateral banks. Here the slope is not smooth.

In general, rivers originate in mountainous regions. Little gutters converge to giverise to a streamlet and the streamlets combine to form a stream at downslopes.

Several streams unite to form a river.

A river is defined as a body of running water carrying sediments which flows along a definite path. The path of the river is the river valley.

Gullies are relatively narrow and deep water courses created by temporary streams resulting from the fall-out of atmospheric precipitations or snow melts.

Apart from the rain water falling on the stream surface or by the melting of snow and glacier ice, usually streams are supplied with water by tributary streams, by seepage from the valley-side slopes by water emerging from underground sources in springs etc. The flow of water in a river or stream is expressed in terms of the volume passing through a point in a given time. This is known as the discharge of the river or stream concerned. It is calculated from measurements in the stream channel, as:—

Discharge = Velocity x Channel cross-section area.

In general, the cross-section area is calculated by multiplying the depth of water in the channel by the width of the occupied channel. The velocity is measured by a current meter at selected intervals across the channel so that the average may be calculated. The velocity of a stream itself depends on the gradient of the channel, the volume of water in the stream, the nature of the channel (i.e. whether smooth, rugged or rough) and the load of sediments in the stream water. Stream velocity also varies along the length of a river, increasing considerably at narrow sections, as compared with wider or deeper sections or pools. All these factors do not reamain constant but change from season to season and from year to year. Accordingly, the discharge of any river or stream is seldom constant.

In rivers the nature of flow of water is characterised mostly by the gradient and velocity. Accordingly, there are two types of flow viz. 1. a laminar or filamented flow and 2. turbulent flow. In the case of laminar flow (also known as streamline flow) the water particles travel in parallel paths. This is possible when the river is having a flat gradient and low velocity. The movement of ground water and also that of glaciers are generally laminar in nature. In the case of natural streams or rivers, the flow of water is so rapid that the motion of water particles is irregular i.e., the rate of flow at each of the river is not constant with regard to either velocity or dirction. Thus the flow is random and eddying. Thus turbulent flow is characterised by eddies, which result in the thorough mixing of the entire mass of flowing water from bottom to top. It has been commonly observed that the maximum turbulance lies just beneath and on either side of the maximum velocity i.e. near the midstream and near the stream where contact between water and bed sets up eddying. When there are eddies the water takes up the debris material from the bottom and brings them in to a suspended state. When a stream follows a curve, then the greatest velocity is shifted towards the deeper water which is generally on the outside of a curve.

A river's work capacity is governed by its kinetic energy. According to the formula of kinetic energy.

$$K = \frac{mV^2}{2}$$, in case of a river

K = the kinetic energy of the river,
m = mass of water or the discharge,
V = velocity of the stream flow.

It indicates that the ability of a river to perform work is more when the river carries much water and the velocity of the stream flow is high or in other words, the rivers work capacity is directly proportional to the mass of water and velocity of stream flow.

GEOLOGICAL-ACTION OF RIVER

Rivers carry excess water from the land to the sea. In doing so, they erode valleys and help in shaping the earth's surface. They transport rock debris and dissolved materials and eventually they deposit most of their sediments in the oceans. River activity in combination with weathering and mass-wasting predominates by far over other types of erosion such as wind, ice or marine. Accordingly rivers are considered as the most important of the geomorphic agents in bringing about the degradation of the land surface.

The geological activity of river is divided chiefly into three parts as:

1. Erosion.
2. Transportation.
3. Deposition.

EROSION

The term 'erosion' is applied for rock breakdown by the dynamic action of any geomorphic agent like moving water, blowing wind, glaciers etc. It is usually defined as the sum total of the process of wearing away of the rocks by the physical forces and chemical factors associated with the natural agencies. Rivers flowing over various rocks break them down purely mechanically. The chemical action of rivers is minimal. A wide variety of processes are included in river erosion as :—

(A) Hydraulic action.

(B) Abrasion.

(C) Attrition.

(D) Cavitation.

(E) Corrosion.

Of the above processes, the first four are the types of mechanical erosion and the last one belongs to the type of chemical erosion. Erosion is mainly responsible for the development of river valleys through river-bed erosion (i.e. down-cutting of the river bed) and lateral erosion. While river-bed erosion is predominant during the initial stages of development of a river, lateral erosion becomes significant during the later stages. It may, however, be remembered that the processes of erosion operate in their own ways, both for down-cutting or lateral cutting of river-valleys.

Hydraulic Action

It is the process of mechanical loosening or removal of the material by the action of the water alone. As we know, it is the turbulent flow of water which can loosen rock and soil particles along the river channel and move them away. Maximum turbulance gives maximum capacity for erosion. It is, therefore, that maximum erosion takes place in the belt of greatest turbulance along the outside of a stream bend. The effectiveness of hydraulic action of a river is dependent on the following factors:

(a) *Gradient* It is the the angle of inclination of two points along the stream divided by the horizontal distance between these two points.

(b) Velocity of the stream.

(c) Width, depth and shape of the channel.

(d) Discharge i.e. the amount of water flowing in the river at a given time.

The stream channel and banks are eroded by moving water due to the forces inherent in them. Hydraulic action's effectiveness is best observed where the flow of water is responsible for undercutting banks of unconsolidated alluvium.

Abrasion

The material which are being carried away by the running water act as tools of destruction and during their transportation, because of their rubbing against the surface of the bed-rock, they bring about a scrapping of the surface. This process of wearing-away of surfaces by mechanical processes such as rubbing, cutting, scratching, grinding, polishing etc. is known as *Abrasion* or *Corrasion*.

Without the presence of rock particles, rivers cannot scratch and scour their channel or in other words, the rock-wastes which are being carried in the river current as load are responsible for the abrasive action of the river. Three types of situations may arise, as follows on the basis of the hardness of the transported material:

(i) if the rock-waste is hard and the bed-rock is soft abrasion of the bed-rock is more pronounced.

(ii) if rock-waste is hard and bed-rock is also hard, it results in the polishing of the bed-rock.

(iii) if rock-waste is soft and bed-rock is hard, abrasion of the bed-rock is not remarkable, since the rock-waste, in such cases, is itself eroded away.

Abrasion may be vertical or lateral. While vertical abrasion causes deepening of the channel, lateral abrasion gives rise to valley-widening. The abrasive action of the river is considered to be the most important means of erosion in bed-rock which are too strong to be affected by simple hydraulic action.

Attrition

The products of abrasion and hydraulic action are carried away in the river-flow, which often collide among themselves, with the bed-rock and also with the sides during their transit and in turn get themselves teared. This process of mechanical wear and tear of the transported rock fragments through which big boulders are gradually reduced in size and finally reach the size-grade of sand and silt, is known as attrition.

Since, in this process rock-fragments in transit suffer mutual collision and repeated impacts, their irregularities and angularities are worn out and they become spherical and rounded.

Cavitation

It is sometimes considered as a type of hydraulic action. It is particularly observed where river water suddenly acquires exceptionally a high velocity such as at water-falls. Rapid increase of velocity reduce the internal pressure in the water. It has now been established that where stream velocity exceeds 12-14 m/sec, the water pressure at that point equals vapour pressure of water and small bubbles of water vapour form and the water foams. As soon as the velocity is decreased due to friction against the floor or sides of the channel, the internal pressure increases again and the bubbles become unstable. The bubbles then collapse suddenly and violently resulting in shock waves which deliver hammer-like blows to the adjoining surface (bed and banks) producing a crop of rock-fragments to be carried away by the river flow. This process, thus, produces hollows in river beds which, in due course, are developed into pot-holes.

Sometimes, highly turbulent rivers in rocky channels erode their beds by hydraulic plucking, in which blocks of bed-rocks etc. are lifted out by suction in strong eddies spiraling up around vertical axes. This sucking out of the rock pieces produces depressions or holes within the rock, which may develop into pot-holes in due course of time.

Thus the rate of erosion is speeded up due to the process of cavitation.

Corrosion

The chemical processes of rock-erosion by river-water are known as *Corrosion* or *Solution*. In the presence of some aiding substances like alkali matter and gases like carbon-dioxide etc, river water is capable of dissolving matter from minerals that constitute the bed-rock in the channel and the rock particles in transport. It has been observed that whenever the water containing carbon-dioxide comes in contact with limestone, it gets dissolved into the water easily. The effectiveness of the process of corrosion depends much on the composition of the rock constituting the river bed (i.e. its solubility) and the composition of the river-water.

IMPORTANT EROSIONAL FEATURES PRODUCED BY RIVER ACTION

1. *Pot-holes* These are cylindrical or bowl-like depressions in the rocky beds of streams, which are excavated in the floors of the streams by extensive, localised abrasion. These are commonly formed in the softer bed-rocks of the stream floor. The drilling of the pot-holes are usually caused by stones swirled round by eddies. The boulders and pebbles etc. act as boring tools and themselves becomes rounded in the process, while the pot-holes continue to grow in size. Pot-holes may vary in dimensions ranging from a few centimetres to several metres.

2. *Water falls* When the river flows over a surface with an originally broken relief, the river literally falls from a height and acquires normal flow against some distance below. Thus, sheer precipices along the course of a river give rise to waterfalls.

As we know, the surface over which a river flows is usually uneven and sometimes the rocks composing the surface are of different hardness. The harder beds may withstand erosion by river, while the softer ones are relatively quickly worn down causing a local difference of elevation in the channel. Two types of situations may arise.

(i) if a hard bed dips gently down the stream, that the steepness is not so pronounced, the river passing over it generally forms a *rapid.* These are water-falls of small-dimensions.

A & B = Outcrop of resistant rocks.
X & Y = Slope on which the river flows.
1,2,3, = Soft beds.

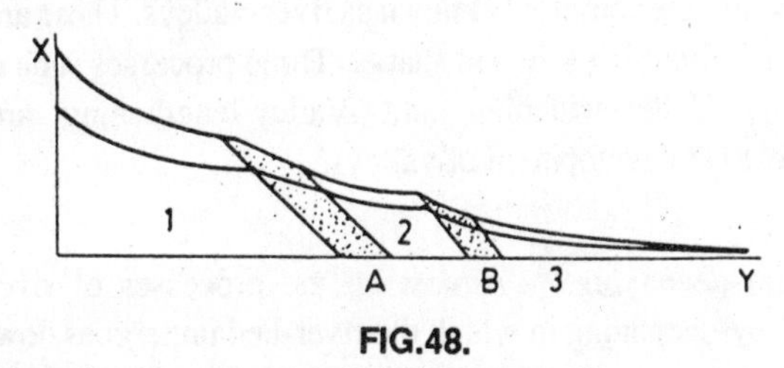

FIG.48.

(ii) if the hard bed is horizontal or dips gently up the stream, the river will erode away partially the softer rocks beneath it. In such cases, the hard rock may stand as a ledge where from water jumps down, falling on lower beds with increased velocity, giving rise to what is known as a waterfall. Thus, when a river falls from a vertical escarpment it forms a waterfall.

Due to recession, gradually there takes place a diminution in height of a waterfall and with continued recession, waterfalls eventually degenerate into rapids and become extinct.

A fall that descends in a series of leaps is referred to as a *Cascade*, Gerosoppa (Jog) water-falls in the Swarvati River in Karnataka is the highest water-fall-in India.

3. *River valleys* The typical river rises in the highlands and flows down due to the pull of gravity. The channels carved out by the flow of

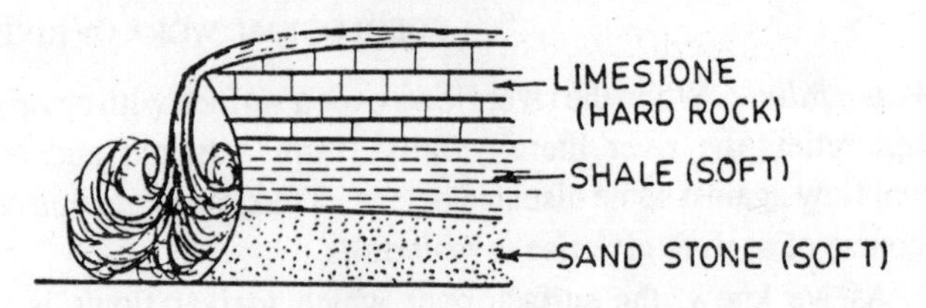

FIG.49. (a) [WATER FALL OVER HORIZONTAL BEDS]

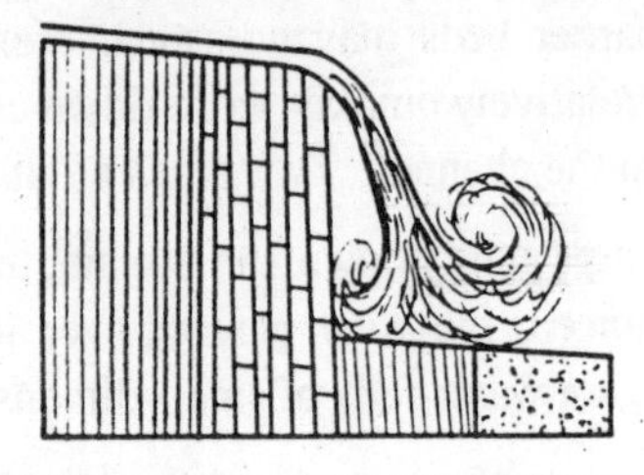

FIG.50. (b) [WATER FALL OVER VERTICAL HARD BEDS]

running water are commonly known as river-valleys. These are negetive land forms of varying size and shape. Three processes such as Valley-deepening, Valley-widening and Valley-lengthening are mainly responsible for development of valleys.

(a) *Valley-deepening* Almost all the processes of river erosion cause valley-deepening in which the river-bed undergoes down-cutting giving rise to a narrow but deep valley. The down-cutting of the valley floor takes place in the upper part of the course of any river, where it flows down the hill slopes i.e. in the mountain or highland tract. This process of valley deepening gives rise to important geological features like gorges or canyons.

Gorges or Canyons

When the river erosion is confined to down-cutting of its channel only, it gives rise to a deep-cut narrow valley, with steep or vertical walls known as Gorge or Canyon, in which the confined water rushes with tremendous force. The Grand Canyon of Colorado river is the greatest

Canyon in the world. It is 900 to 1800 metres deep, 60 to 90 metres wide and extends for a length of 300 kilometres. Deep gorges are found in mountainous areas and plateaus where the rocks constituting the river-bed are chemically resistant and mechanically strong.

The process of down-cutting of the valley-floor does not continue for ever. Gradually the rate of deepening slows down and commonly it stops when the *base-level of erosion* is reached at a late stage. This happens when a river completes its course i.e. meets the sea or lake, whereby it loses its erosive power.

The *Base-level of erosion* of a river is the level of the basin in to which it falls and at the level of which the river loses its kinetic energy and therefore a river cannot excavate its channel below this level. It should be noted that when a river falls into the sea or ocean the sea-level must be taken as the base-level; when it falls into a lake, the base-level is the level of the water in the lake. Base-level of erosion is commonly defined as the mean-sea-level produced inland by river erosion.

(b) *Valley-widening* Lower down the highland tract there is a gradual reduction in the channel gradient of a river and the erosive power of the river to cut downwards becomes less; but the river starts cutting sideways with wide swinging curves and meanders.

A number of processes may be attributed to the phenomenon of valley-widening which are as follows.

(i) *Lateral erosion* Through erosion of the valley sides mostly by hydraulic action and abrasion as well as through slumping of the materials in to the river because of undercutting of the valley-walls, the net effect of which is to broaden the channel.

(ii) Apart from the process of lateral erosion, the processes of rain-wash (or sheet wash), gullying, weathering and mass-wasting and incoming tributaries etc. lead to the widening of valleys.

(c) *Valley-lengthening* Lengthening of river-valley is usually achived by the process of headward erosion (or regressive erosion), where the long profile of the river develops from the base-level towards its sources. This can be illustrated as follows:

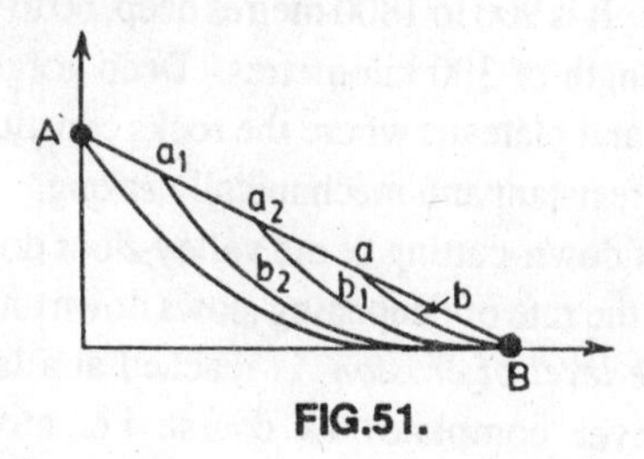

FIG.51.

In the given figure,

A = Source of the river.

B = Base-level of the river.

AB = the surface along which the river had flowed initially.

It is assumed that along the surface AB, the points a_2 a_1 and a represent various tributaries falling into the river. As such, there will be more water for discharge at point a (because here the river receives the water from a number of tributatries also) and erosion will be more intensive. Thus the river bed in the aB sector is strongly cut and there develops a steep slope occupying the position abB. This enhances the velocity of the current and the bottom erosion above point a gets intensified with the result that the channel deepens on sector aa_1 and the river occupies the position a, b, bB. The same process is repeated till the river valley is excavated to such an extent that instead of the original position the new smooth curve is in the shape of a parabola. Thus a backward extension or lengthening of the river-valley takes place.

Apart from the above process of headward erosion, the other processes responsible for lengthening of river valley are as follows:

(i) through increase in the size of their meanders;

(ii) uplift of the land or lowering of the sea-level results in extension of the valley through the newly exposed land etc.

Development of the Shape of the River-Valley

In the upper part of the river's course i.e. in the mountainous and hilly tracts, where bed-rock erosion is maximum, the channel deepens and forms a valley with vertical sides. But with the aid of other agencies, the sides of the valley are also cut and the valley becomes V-shaped.

If the rocks, in which the valley is cut, are hard and resistant or porous, the sides will be steep and the valley will be narrow whereas if the rocks are soft and can be easily worn away, the valley will form a wide and open 'V'. Besides, the 'V' widens out as the valley reaches maturity more and more, where lateral erosion dominates and gets the valley floor more and more flattened as shown in the diagrams given below:

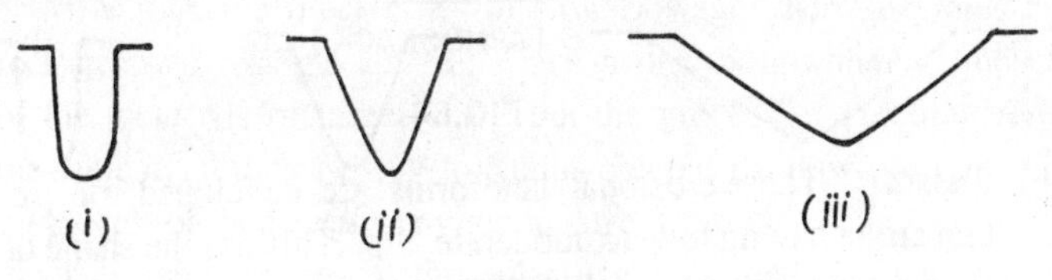

FIG.52.

4. *Escarpments* These are erosional land forms produced by rivers in regions composed of alternating beds of hard and soft rocks. The differential erosion of rocks give rise to a steep slope, called escarpment. This is usually developed in dipping beds with harder rocks overlying the soft ones. Such structures are commonly referred to as homoclinal structures. Because of the variation in resistance to erosion, the hard rocks are eroded at a very slow rate whereas the soft ones are eroded comparatively at a rapid rate. This gives rise to a steep slope on one side of the structure and a gentle slope on the other. The steep side is known as escarpment or scarp face.

There is a number of similar erosional features as follows:

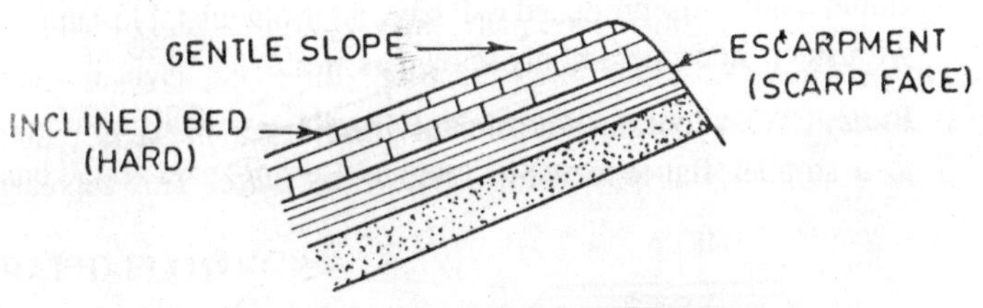

FIG. 53. [ESCARPMENT]

(a) *Hogbacks* These are sharp crested often sawtooth ridge formed of the upturned edge of a resistant rock layer of sand-stone, limestone or lava. In these cases, the beds dip at a high angle, roughly in excess of 45°, so that the dip slope becomes almost as steep as the escarpment. Thus the ridges have steep slopes on both the sides, as may be seen in the following diagram

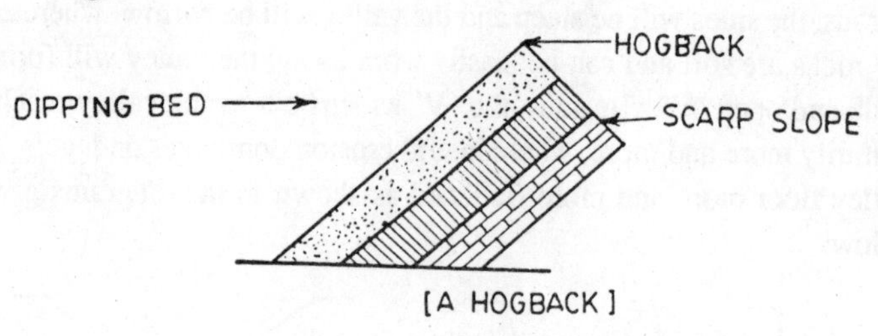

FIG.54.

(b) *Cuestas* These erosional landforms are developed on resistant strata having low to moderate dip. This has the shape of an asymmetric low ridge or hill belt with a steep scarp on one side and a gentle slope on the other. Majority of the cuestas are associated with coastal plains.

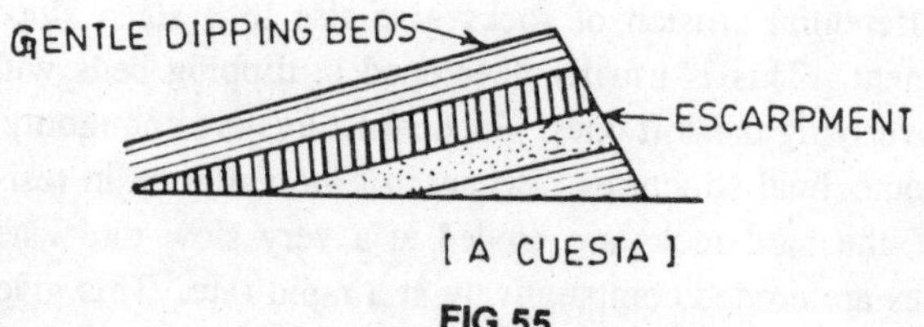

FIG.55.

(c) *Mesa* In regions of horizontal strata in which isolated portion of land is capped by a hard, erosion-resistant bed, the erosional landforms produced will have an isolated table-land area with steep sides, commonly known as *mesa.*

(d) *Butte* With continued erosion of the sides a mesa is reduced to a smaller flat-topped hill, known as *butte*. This represents

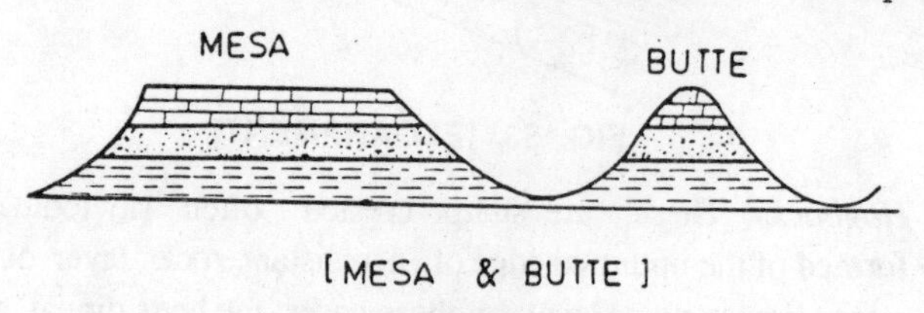

FIG.56

the final remnant of a resistant rock layer in a region of horizontal strata. Such features in South Africa are called *Kopjes.*

5. *Peneplain* Peneplain is the term used for the nearly smooth erosion surface of relatively low relief and altitude which covers a large area. This is a landform which is the ultimate of the erosion cycle and it evolves in a temperate humid climate. Here the slope is so gentle that the river's velocity down the slope is not sufficient to move the products of erosion and weathering.

The surface of the peneplains may be dotted here and there by a relatively few, small rounded hills, underlain by resistant rocks. These hillocks made up of harder, durable and resistant rocks on the surface of the peneplains are termed as *Monadnocks*; which are the remnants of the pre-existing country which could survive the vehemence of fluvial erosion.

6. *Pediments* These are the erosional landforms mostly found in the arid regions. The landform is a gently sloping plain worn down by scarp retreat. They are usually located in the piedmont area, fringing an upland massif, but sometimes in wider low land zones. They may or may not be covered by a thin veneer of alluvium but scattered over it are rock fragments,some brought by running water from the adjacent mountains and some derived by weathering from those immediately beneath. Retreat of the mountain slope lengthens the pediment at its upslope edge.

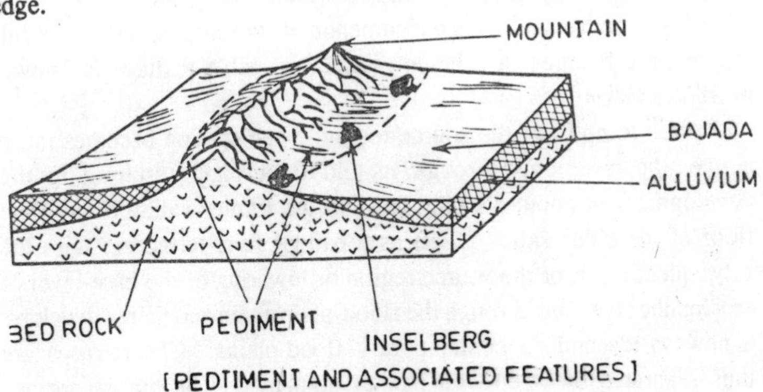

[PEDTIMENT AND ASSOCIATED FEATURES]

FIG.57.

The vast plains formed by the convergence of many pediments are termed as *Pediplains*. It consists of areas of pediment merging with surrounding alluvial fan and playa surfaces. Pediplains are often considered as analogous to peneplains.

Isolated concave-shaped masses of rock rising above a pediment or pediplain as round-topped mounds are known as *Inselbergs* (German term meaning *island mountain*). These are entirely analogous to monadnocks occurring on a peneplain and appear near the ultimate stage in the erosional cycle. These residual hills are also known as *Bornhardts*. Typical bornhardts are prominent knobs of massive granite or similar plutonic rock with rounded summit and often showing exfoliation shells.

7. *Wadies* These are the channels formed during rains in desert or arid regions and are also known as *washes*.

8. *River-terraces* These are erosional features consisting of several step-like plains along the side of a river valley. As we know, in the absence of tectonic disturbances, the river attains a profile of equilibrium in due course of time and there will be a marked growth of lateral erosion and deposition; the down-cutting of the valley gets stopped and the valley becomes nearly a plain. But with the subsequent revival of tectonic movements when the source region is uplifted the gradient of the river increases resulting in the increase of the velocity of the river. This will again start down-cutting of the valley floor and development of topographic features which are characteristic of a river in its youth stage. This phenomenon of development of youthful topographic features in a landmass having stable features is known as *Rejuvenation*.

Due to uplift of the source region, when erosion becomes more active, the river cuts through its old flood-plain resulting in the development of another valley beneath the former valley floor. The floor of the older valley is left as a pair of river-terraces. With the subsequent uplift of the source region or lowering of the base-level of erosion the river cuts through the flood-plain again and there develops a new valley and a contemporary flood-plains. The terraces are thus separated by steep-wall like escarpments. The highest terrace above the flood plain will, accordingly, be the oldest while the lowest will be youngest. The number of terraces also reflect the number of periods of rejuvenation which have affected the river.

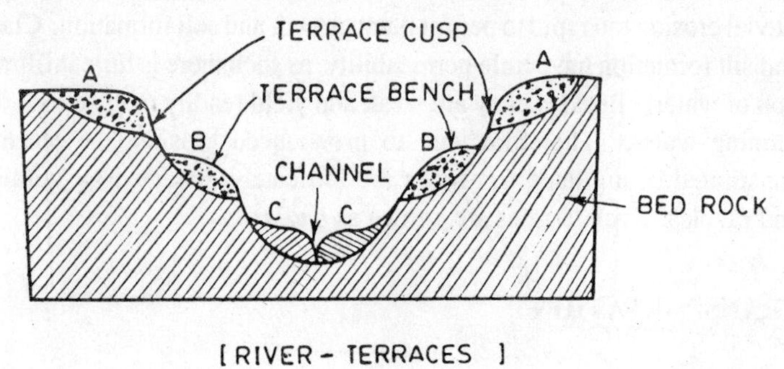

[RIVER – TERRACES]

FIG.58.

There are three types of river-terraces as:

(a) *erosion terraces*, (b) *accumulative terraces* (c) *composite terraces*. Erosion terraces are composed almost entirely of bed rock and little or no alluvial veneer. Accumulative terraces are composed of alluvial deposits. Composite terraces consist of the terrace-bench composed of alluvium while the bed-rock lies at the base of of their cusp. The erosion terraces are also known as *bed-rock terraces*, whereas the accumulative terraces are also known as *alluvial-terraces* and the composite terraces as *rock-defended* terraces.

River-terraces are also classified as (i) Cyclic terraces and (ii) Non-cyclic terraces. In the cyclic terraces, the remnants on the opposite sides of the valley are paired whereas the non-paired terraces are characteristic of non-cyclic terraces.

Apart from giving rise to river terraces, rapid uplift of the land gives rise to *incised* or *entrenched meanders* and *natural rock-bridges*. The meandering course of the river is still maintained but there is rapid deepening of the valley by the increased erosion, consequent on the uplift. As a result, the winding gorge digs deep into the soild rock beneath. Such gorges are called incised *meanders*.

If the neck of an incised meander is narrow, the intervening ridge of rock may be cut through down below from both the sides forming ultimately, what is called a *natural bridge*.

9. *Badlands* These are rugged land surface of steep slopes, which are developed on weak clay formations of clay-rich regolith by fluvial erosion too rapid to permit plant growth and soil formation. Clay and silt formation have little permeability, as such there is little infiltration of water. Besides, they are weak and yield readily to the attack of running water. Therefore, due to pronounced erosion, the terrain constituted of argillaceous rocks are intricately dissected by gullies and ravines. Such terrains are known as *badlands.*

TRANSPORTATION

As we know, the work carried out by a river is related to the energy available to it. Each river has a certain quantity of potential energy determined by the height of its source region and the volume of water entering the river system. This energy is converted to kinetic energy as it moves through the system. The products both of fluvial erosion and weathering constitute the load of the river which are carried downstream along with the flow of running water. That is to say that river transports its load due to the kinetic energy associated with its flow.

The load carried by a river can be subdivided as follows:

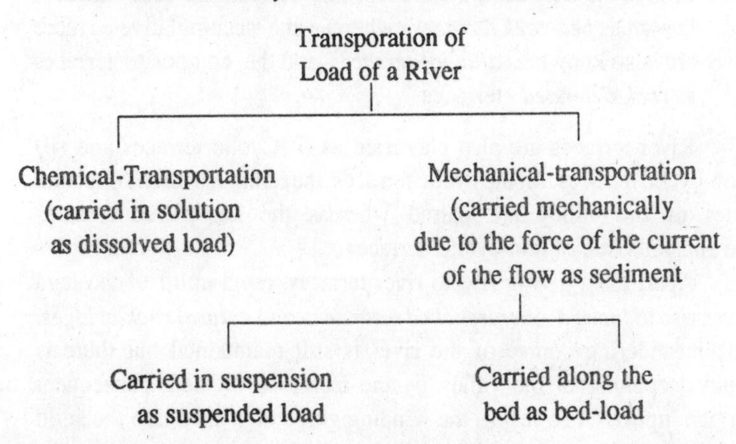

The percentage of contribution to the total load made by the aforesaid classes vary widely with the nature of the river, the climate, the lithology of the river-bed etc. A river carries the greatest amount of materials during floods and spates.

1. *Chemical-Transportation* A considerable amount of mineral matter is tranported in a dissolved state i.e. in solution. The load in solution actually becomes a part of the liquid in the stream. Limestone, dolomite and salts are very much soluble and some compounds of iron, manganese, phosphorous etc. are transported in the form of collidal solution.

2. *Mechanical Transportation* This is influenced by three main factors as follows:

(a) Velocity of the river-current;

(b) Nature of the river-current; and

(c) Density of rock materials to be transported.

The load carried through the mechanism of mechanical transportation are grouped in to two categories as *Suspended load* and *Bed load.*

Suspended Load

Finė particles of clay and silt are transported in suspension. These particles sometime float on the surface of the water and sometime they are carried within the water and become part of the fluid mass. The intensíty of turbulance and the velocity of the river-flow determines the length of time a particle remains in suspension.

Bad Load

It comprises the heavier particles of sand, pebbles, gravels and cobbles which move close to the channel floor by rolling or sliding and an occasional low leap. It is moved along on, and supported by the bed of the channel. The transportation of the bed-load takes place in two ways: (i) *Saltation* and (ii) *Traction.*

(i) *Saltation* Here a particle resting on the river-bed is temporarily lifted up by the eddies and is carried to some distance before it again falls to rest. The smaller the particle, the higher the lift and the longer the jump. This process of saltation is mostly effective in the transportation of sand grains.

(ii) *Traction* The transportation of sediments by creeping is known as 'traction'. Bigger fragments of rocks like boulders, cobbles and pebbles are rolled along the bottom or they may slip and slide downstream. The weight of fragments that can be rolled along the bottom is proportional to the sixth power of steam velocity, according to the Airy's law. Thus a mountain river transports rock-fragments of bigger size much more in comparison to that of the flat-country rivers.

DEPOSITION

Deposition of the transported materials takes place where the river's capacity or transporting ability is reduced. It usually occurs when the following conditions are met

(a) Decrease in the velocity of the river,

(b) Decrease in slope or gradient;

(c) Decrease in volume of water i.e. discharge,

(d) Change in channels,

(e) Chemical precipitation, etc.

All the above factors diminish the velocity of the river and influence the deposition of sediments carried by it. Deposits, thus formed, are called fluvial deposits. The following are some of the important depositional features associated with river action.

IMPORTANT DEPOSITIONAL FEATURES PRODUCED BY RIVER ACTION

1. *Alluvial fans and Cones* When streams flow abruptly from steeper to gentler gradients, as at the base of a mountain or ridge, its velocity is checked and the huge quantities of material carried by the river are dropped there giving rise to a broad, low cone-shaped deposit called an alluvial fan. Thus alluvial fans form where a stream leaves a confined valley and enters a flatter region. The material constituting a fan includes coarse boulders and pebbles at its head to finer material down its slope.

The term alluvial fan is commonly used when the slope of the deposit is below 10 degrees and alluvial cone when the slope is from 10 to 50 degrees.

A series of adjacent fans may in time coalesce to form an extensive piedmont alluvial plain, also known as 'Bajada'.

2. *Flood-plain deposits* Flood plains are areas of low and relatively flat land bordering the channel on one or both the sides, at bank level. These areas are readily submerged under water during flood time, when the river water overtops the banks of the channel and rises above the channel at low water. Deposits formed on flood plain by flood-water outside the actual channel are known as *Overbank deposits.*

A number of features are associated with the flood plains, which are as follows:

(a) *Meanders and oxbow lake* Dominating the flood-plain is the meandering river channel, i.e. the river flows across the flood-plain in broad sweeping curves, known as meanders. Meanders are common where the gradient of a river becomes extremely low. In a flood plain a slight obstacle or accidental irregularity usually causes a deviation of the current with the initiation of a bend. Once started the bends tend to grow and gradually become more pronounced. The water flows faster around the outerside of the bend and is slow on the inner curve. Accordingly, erosion becomes more towards the outside of each bend and the channel deepens along the downstream part of the bend which is also termed as the *under-cut side.* At the same time there starts deposition towards the innerside of each bend forming what is known as the *slip-off slope.* Thus the river shifts its channel towards the outer bank and leaves gently rounded slip-off slope on the inside of the growing curve.

A meander grows until it becomes bulb-shaped with a narrow neck, because of the constant broadening of a river's bend during the erosion of the outer bank and deposition on the inner bank. During floods, the increased power of the flow may carry the stream across this neck. As a result, the river straightens its channel. The former meander, therefore, remains as a back-water for sometime and the entrances to it gradually get silted up since the river follows the shortest course and the water in the meander is still. Thus a meander loop gets abandoned. This phenomenon is called a cut off. The abandoned channel thus constitute a loop-shaped lake known as an *oxbow lake* or *horse-shoe lake*.

The deposits formed at the slip off slope of a meandering river is known as *Point-bar*.

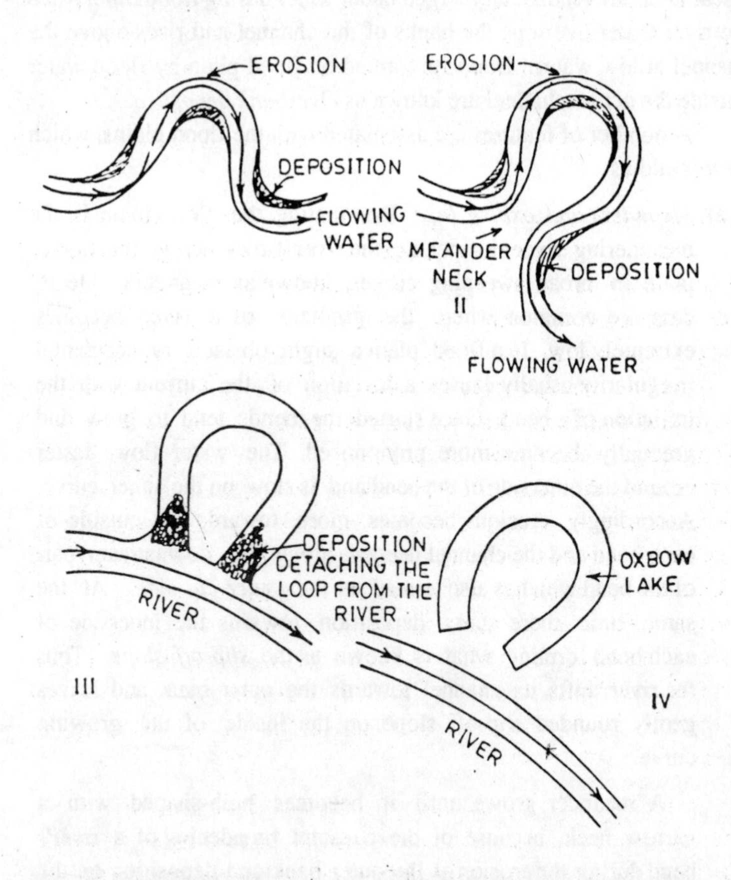

FIG.59. [RIVER MEANDERING AND DEVELOPMENT OF OXBOW LAKE]

(b) *Natural levees* These are broad, low-ridges formed along the bank of the river during floods. During floods, when the entire flood plain is inundated, water spreads from the main channel over adjacent flood plain deposits. When the flood retreats, sand and silt etc. are deposited in a zone adjacent to the channel forming low ridges that parrellel a river course. They are highest near the river bank and gradually slope away from it, because the deposition is more nearer to the channel and decreases away from it.

2. *Braided River* Braiding is a phenomenon of dividing and reuniting the river channels. In such cases, the river flows in a number of narrower channels separated by lenticular sand and gravel bars which may again meet the main channel somewhere downstream. They are

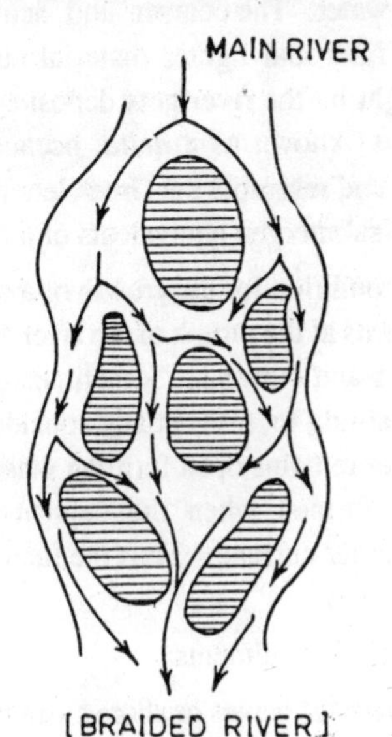

[BRAIDED RIVER]

FIG.60.

commonly developed where the amount of load is more and the river is incapable of transporting all of it. This deposition starts near the centre of the channel. The coarser fractions of the load tend to form islands with a channel on each side and similarly other islands also develop. Accordingly the flow is divided into multiple branches (which may rejoin later on) to give rise to what is known as braided river.

Apart from the above, there are also other causes for the development of a braided-river. Sometimes a considerable portion of water is lost due to evaporation and infiltration, thereby depositing the load of silt and clay carried by the river on the channel itself. This makes the channel so shallow that the stream cannot be contained in it and therefore spills out over one side. This overflowing water is capable of cutting a new track gradually. With subsequent repetition of this process a braided river pattern is developed.

3. *Delta* Deltas are basically features of river deposition. As we know, when a river enters a lake or sea its velocity is checked rapidly and the process of deposition is accelerated. Even the colloids carried in the river water get coagulated due to the electrolytes present in the sea water. The coarser and heavier material is laid down first and the finer and lighter material is carried further out. Thus the load brought by the river gets deposited at its mouth, which gives rise to what is known as a *delta*; because these deposits are triangular in outline and resemble the Greek letter Δ (delta). Deltas are considered to be the submerged equivalents of alluvial fans.

The essential condition for the growth of a delta is that the rate of deposition of sediments at the mouth of the river should exceed the rate of removal by waves and currents. Sometimes the tides and currents may be sufficiently strong to prevent any considerable deposition and the mouth of the river remains open forming what is called an estuary; whereas deltas are formed when the deposits of a river are not removal by tidal or other currents. Thus the factors favourable to the formation of a delta are :

(a) Abundant supply of sediments;

(b) Absence of powerful waves or shore currents;

(c) A stable body of water;

(d) A shallow water offshore;

Small deltas may exhibit a characteristic pattern of stratification not present in many large deltas-built into the ocean. Thicker layers of coarser-grained sediments known as *foreset beds* pile up on the sloping bottom close to shore, whereas finer sediments deposited in thinner layers further out are known as *bottom-set beds*. The bottom-set beds are actually the continuations of the foreset beds. On top of foreset beds, thin layers of sediments lie, which have a gentle seaward slope. These are known as *top-set beds*

Deltas show a variety of shapes, mostly because of the configuration of the coastline as well as the action of the sea-waves. On the basis of the shape the deltas are classified as—arcuate delta, bird-foot delta, cuspate delta etc.

Through the delta run a large number of channels which come out of the main channel. The smaller channels are termed as distributaries.

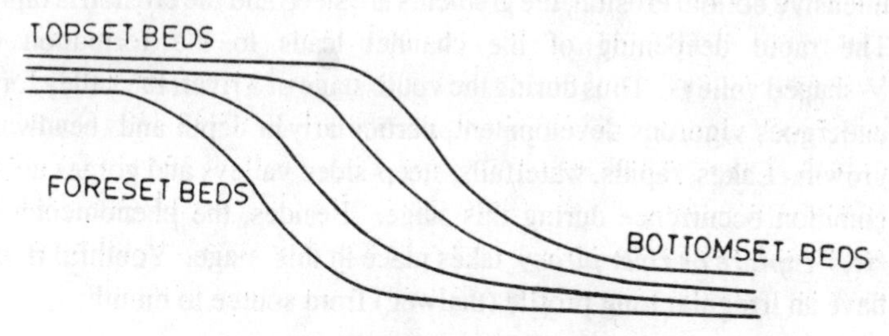

FIG.61. [BEDDING OF A DELTA]

Deltas provide extensive flat fertile lands which support dense agricultural population, as for example, the Ganges delta, the Nile delta, the Mississipi delta etc.

CYCLE OF EROSION

The concept of cycle of erosion was formulated by William Morris Davis, an American geomorphologist, towards the end of the nineteenth century. It is a concept of an orderly sequence of evolutionary stages of fluvial erosion in which relief of the available landmass declines with time to reach a late stage when the landscape becomes a peneplain . The cycle of erosion, as envisioned by Davis, has its initial stage at a time when the landmass is rapidly elevated by internal earth forces, followed by a very long period of tectonic quiescence. Once raised high above sea level as a landmass, streams come into existence and erosion begins to operate on the uplifted mass which is gradually worn down almost to a plain. The landmass may, at some later time, be rejuvenated and the cycle begins again and remnants of the earlier cycle of erosion are preserved at new and higher levels.

In a normal cycle three stages have been recognised as: *youth stage*, *mature stage* and *old stage*. These follow each other in a regular sequence.

Youth Stage

In this stage the river flows along an uneven surface and there is intensive bottom erosion, the gradients are steep and the erosion is rapid. The rapid deepening of the channel leads to the formation of V-shaped valleys. Thus during the youth stage of a river, the valley form undergoes vigorous development, particularly in depth and headward growth. Lakes, rapids, waterfalls, steep-sided valleys and gorges are of common occurrence during this stage. Besides, the phenomenon of *river-capture or river piracy* takes place in this stage. Youthful rivers have an irregular long profile (thalweg) from source to mouth.

River Capture

When one of the two rivers flowing in opposite directions from a single divide, becomes more effective in erosion due to steeper gradient (when the slopes are unequally inclined), the divide gradually recedes towards the side with the gentler slope. In other words, the river with steeper gradient extends its valley headward thus causing a shift of the divide against the river with gentle gradient.

Gradually deepening of the valley continues headward with pronounced dissection of the ridge (divide). Sometimes this headward migration of one river enables it to reach the river on the other side. But, as the first river has a steeper gradient than the other one, the course of the second river gets diverted and its water starts draining through the channel of the first river. This process of diversion of a river by the headward migration of another river is known as *River-Capture or River-piracy.*

The point where the course of the second river is diverted is known as the *Elbow of capture*. The captured river is known as *Misfit* and the deserted part of its channel through which no water flows is termed as the *Wind-gap*.

Mature Stage

In this stage rivers flow with a graded profile i.e. it attains a profile of equilibrium. The land mass is fully dissected and a well-integrated drainage system is developed. Ridges and valleys develop prominently. Flood plains develop and river meandering takes place. The topography consists of features such as: hogbacks, cuestas, mesa, butte, meanders, oxbow lakes, natural bridge, flood plains, alluvial fans etc.

Old Stage

In this stage the gradients are gentle and the velocity is low. Accordingly the river lose most of its erosive power and flow in a sluggish manner. In old age a river has maximum meandering. The river at this age does little of erosion and transportation but is mostly engaged in deposition. This stage is characterised by the development of distributaries and the river flows almost at the base level of erosion.

The topography consists of features like peneplains, natural levees, deltas etc.

Most of the cycles of erosion do not reach the final stage, as sometime during their operation either climatic or tectonic disturbances take place, and thus results in an incomplete or partial cycle.

RIVER PATTERNS

On the basis of their characteristic features, origin and manner of development, rivers are classified into the following types as:—

(a) *Antecedant river* : Such rivers exist before the surface relief was impressed upon the area. eg. Brahmaputra river.

(b) *Consequent river* : These rivers flow as a consequence to the existing surface relief.

(c) *Subsequent river* : The river that joins the consequent river, arising later as erosion proceeds.

(d) *Insequent river* : It does not indicate any particular reason for its pattern and course of flow such as that upon homogeneous terrain.

(e) *Obsequent* : Here the river drains in the opposite direction to the original consequent river.

(f) *Resequent* : Such rivers drain in the same direction as the original consequent river, but at a lower topographical level.

(g) *Super-imposed* : These rivers are independent of the geological structure of the country through which they flow. This can be explained as follows:—

At some places, old rocks may be covered under a sheet of new deposits. Any river developed on such an area will follow the surface relief of the overlying cover and will not have any relation with the older rocks lying below. Gradual erosion removes the overlying cover and then the river flows on the older rocks below. In such cases, the river is said to be super-imposed on the older rocks below.

DRAINAGE PATTERN

The joining of the tributaries with the master stream produces a pattern termed *drainage pattern*. This is largely a reflection of the effect of structure on streams. The most common drainage patterns are as follows:

(i) *Dendritic* : It is characterised by irregular branching of tributary streams, in a similar pattern as that of a tree's branches. This is characteristic of flat regions with broadly uniform composition.

(ii) *Parallel pattern* : It develops on streep slopes where the tributaries and the master stream flow parallel to each other.

(iii) *Trellis pattern* : It is a rectangular drainage pattern which develops when a consequent stream receives a number of subsequent streams from either side at approximately right angles. It i s mostly common in a topography created on a folded structure of synclines, anticlines, faults or joints.

(iv) *Radial pattern* : It consists of drainage lines radiating from a central part as on a dome.

RIVERS IN INDIA

The rivers in India are classified as:

(i) Himalayan rivers;

(ii) Peninsular rivers;

(iii) Coastal rivers;

(iv) Rivers of the inland drainage basin.

The Himalayan rivers are snow fed and have reasonable flow throughout the year. The Ganges system, the Brahmaputra system and the Indus system are the three major systems of Himalayan river which include about 23 principal rivrs like Jamuna, Manas, Sutlej etc.

The Peninsular rivers are generally rain-fed and therefore fluctuate in volume. They are more or less completely graded. The western ghats is the main watershed in the Peninsula. Major rivers like Mahanadi, Godavari, Krishna, Kaveri etc. flows eastwards and drain in to the Bay of Bengal. Narmada and Tapti are the important west flowing rivers which drain into the Arabian Sea.

The coastal streams, specially of the West coast are short in length and have limited catchment areas. Most of them are flashy and non-perennial.

The streams of the inland drainage basin of western Rajsthan are few and far between. Most of them are of an ephemeral character. They drain towards the individual basins or salt lakes like the Sambhar or are lost in the sands, having no outlet to the sea. The Luni is the only river of this category that drains into the Rann of Cutch.

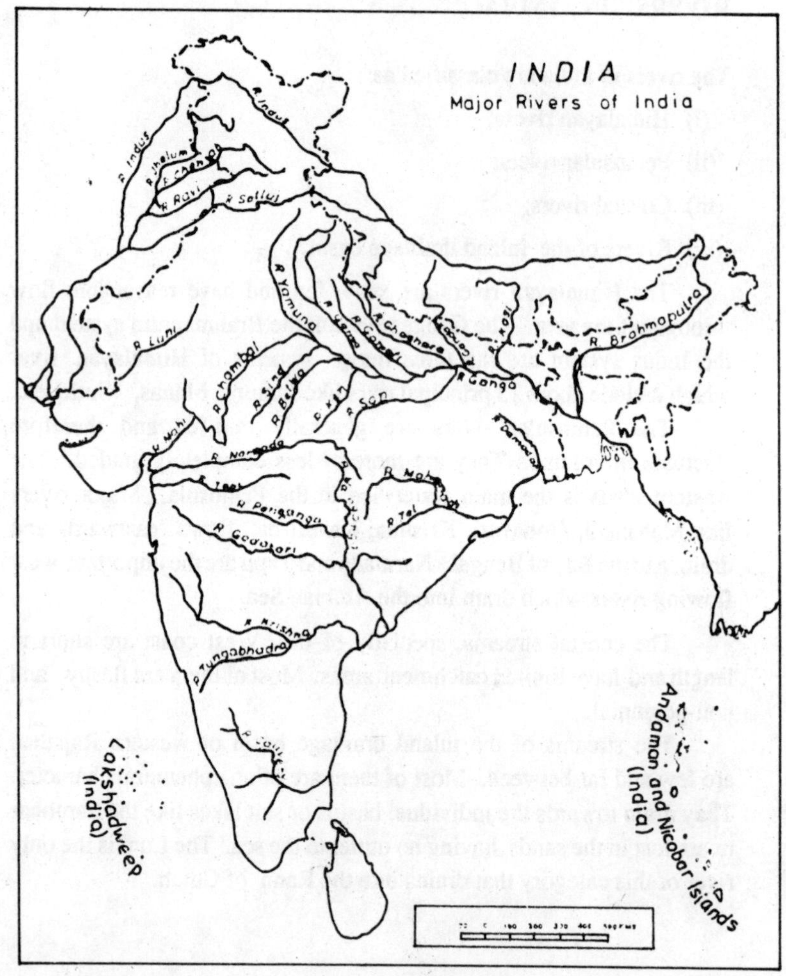

Based upon Survey of India map with permission of the Surveyor General of India. The territorial waters of India extend into the sea to a distance of twelve nautical miles measured from the

FIG.62. Major Rivers of India

22

GEOLOGICAL WORK OF WIND

Wind is the moving air. Wind blowing over the solid surface of the lands is also an active agent of landform development. Its activity is particularly intensive in the deserts and semideserts which constitute about 20% of the surface of continents. The geological action of wind is particularly effective in areas that lack plant cover, have a considerable diurnal and seasonal temperature variation, and low precipitation.

The geological action of wind can conveniently be divided in to three stages viz. Erosion, Transportation and Deposition. As a whole, the geological action of wind is largely governed by its velocity. But wind alone has little influence on shaping the surface of the ground, because it is only able to move small dry particles. In humid climatic regions, the surface of the earth is protected by a solid cover of vegetation and also by the cohesive effects of moisture in the soil from sharp temperature fluctuations causing physical weathering and the deflation work of the wind.

EROSION

Wind erosion manifests itself in three forms viz. (i) deflation, (ii) abrasion or corrasion and (iii) attrition. Wind uses sand as the agent of erosion. Wind and running water are in many respect similar in the ways in which they erode and transport sediment particles.

Deflation

A strong wind can transport very coarse sand, lifting it from the ground and carry it for great distances. This process of removal of loose soil of rock particles, along the course of the blowing wind is known as 'deflation' (from the Latin de flare = to blow off). The wind picks up and removes loose particles from the earth's surface, and thus helps to lower the general level. This process operates well in dry regions with little or no rainfall. The rate of deflation depends on the force of the wind, the nature of the rock and the degree of weathering it has suffered etc.

Features Produced by Deflation

(i) *Hamada* When the loose particles are swept away the hard mantle left behind is known as 'hamada'. The term has been applied to the stone-strewn surface in the Sahara desert, left after the finer materials are removed by wind. This is a form of lag-deposits.

(ii) *Blow-outs or deflation-hollows* Deflation sometimes leads to the formation of depression or hollows on the land surface. At few places, deflation may continue to deepen a blow-out in fine-grained sediment until it reaches the water-table. These depressions may range from a few metres to a kilometre or more in diameter, but it is usually only a few metres deep.

Such depressions, when deepen until the water-table and gets filled with water, create shallow ponds or lakes known as 'Oases'. The position of the oasis is quickly stabilised by the growth of vegetation-commonly palm trees. Some oases are very small with only a few trees, whilst others are large enough to support moderate-sized townships surrounded by gardens and date palms. The *pans* of South Africa, the so-called lakes of west and central Australia etc. are probably the results of long-continued deflation.

(iii) *Lag deposits* Sometimes a layer of residual pebbles and cobbles are strewn upon the surface while intervening finer particles have been removed as a result of deflation. These accumulations of pebbles and boulders have been designated by the general term *lag-deposits.*

By rolling or jostling about, as the finer particles are removed, the pebbles become closely fitted together forming what is known as a desert pavement. This layer protects the underlying sediment from further deflation. Its widespread occurrence is emphasised by the variety of names applied to it : *reg* in Algeria; *rig* in Iran, *serir* in Libya; *the gibbers* in Australia.

Abrasion

The loose particles that are blown away by the wind serve as tools of destruction, wearing away the surface with which it comes in contact. This process is also known as *corrasion.* Abrasion is mainly effective as part of saltation (a mode of wind transport) and can operate only near the ground because of the inability of wind to lift sand more than a few feet. Its main effect is mostly seen in under cutting and fluting at the base of upstanding rockmasses. Depending on the hardness of the rock and the character of the material borne by the wind, the surface of rocks is polished, covered with striations, furrows or grooves, and so on. For effective abrasion, sand-blasting must continue for a long time and the wind must have a long fetch across a source area of suitably-sized particles.

Features Produced by Abrasion

(i) *Yardang* It is a grooved or furrowed topographic form produced by wind abrasion. The grooves are elongated in the direction of prevailing winds and are separated by sharp ridges.

The yardangs commonly develop, where the exposed rocks have vertical layers, consisting of alternations of hard and soft strata, and when the winds are steady and blow in one direction, the softer strata are scoured away more rapidly than the hard and resistant strata. Thus, there develops a topographic feature consisting of elongated ridges and furrows, depending on the original rock characteristics. These are also usually under-cut. These are common in parts of the Asiatic deserts.

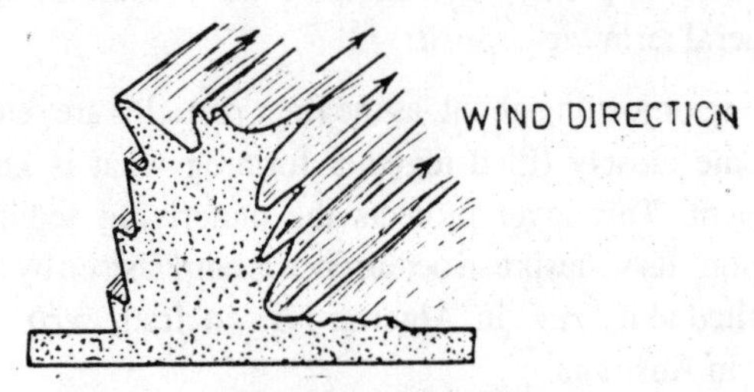

FIG.63. (A YARDANG)

(ii) *Ventifacts* These are the pebbles faceted by the abrasive effects of wind-blown sand. These are developed when sand has been blown over pebbles for a longtime, so that they become worn from the repeated abrasion and smooth polished surfaces result. Ventifacts with one smooth surface is called Einkanters, with two abraded surfaces as Zweikanter and with three smooth faces as *Dreikanters*.

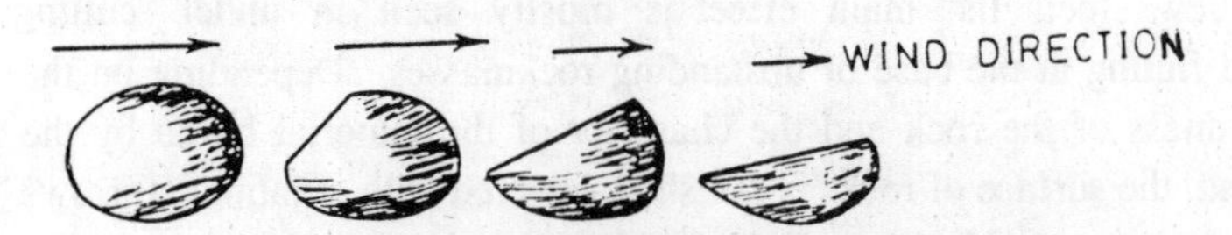

FIG.64. (DEVELOPMENT OF A VENTIFACT)

(iii) *Pedestal rock* It is a wide rock-cap standing on a slender rock column, produced because of wind abrasion. As we know, the sand-blast action is most effective just above the surface of the ground where the drift is thickest and it decreases rapidly upwards as a result of which rocks which projects upwards are under-cut. When soft rocks capped with harder and resistant rocks are exposed to wind abrasion, the softer rocks being more deeply worn, produce a mushroom-shaped form in which the upper widened part of the rock rests upon a relatively thin and short rock-column.

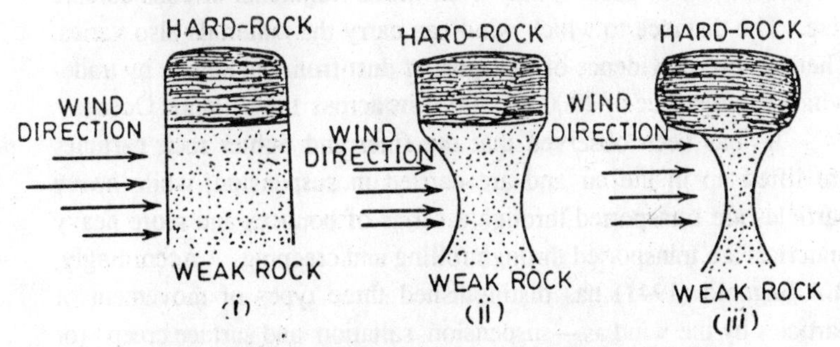

FIG.65. (DEVELOPMENT OF A PEDESTAL- ROCK)

(iv) *Zeugen* These are tabular masses of more resistant rock resting on under-cut pillars of softer material and are very often elongated in the direction of prevailing wind; besides the strata are horizontal.

Attrition

While on transit, wind born particles often collide with one another and such mutual collision brings about some degree of grinding of the particles. Thus rounding of grains become perfect to a great extent and the grains are reduced to smaller dimensions. The more the length of transit and velocity, the greater is the degree of rounding.

Feautures Produced by Attrition

Millet-seed sand These are rounded deserts and grains, produced through the process of attrition and have resemblance with millet-seed grains. Sands of this type are seen sometimes in ancient formations giving indication about the former presence of deserts.

TRANSPORTATION

Wind is an active agent of transport of fine materials, especially sand and dust, and moves thousands of tons of these materials from one place to another. Wind-transportation is dependent mainly on wind-velocity. As such, during a gale the wind often blows the smaller pebbles along the beach, and in gusts it may even move fragments of considerable size. The distance to which wind can carry the materials also varies. There is, even, evidence of transport of dust from the Sahara by trade-winds to a distance of 2000 to 2500 Kms across the Atlantic Ocean.

It has been observed that the finer and lighter rock particles are lifted up in the air and are carried in suspension, while heavy particles are transported through a series of bounces and more heavy materials are transported through rolling and creeping. Accordingly, R.A.Bagnold (1941) has distinguished three types of movement of particles by the wind as—suspension, saltation and surface creep (or traction).

Suspension

As we all know, the wind can carry with it any loose material which lies upon the surface and which is sufficiently light and fine. It has been observed, that the finest material, derived from silts and clays and having a diameter of less than 0.05 mm like dust cloud, smoke etc. move with the wind quickly and remain in suspension in the air for quite some time and settle very slowly.

Saltation

Medium sized particles having diameters between 0.05 mm to 2.0 mm are moved in a leaping manner. These particles are commonly too heavy to remain in suspension and lighter to be transported through rolling. In a turbulent flow of air near the surface, the wind initially may cause a particle to roll forward and, on knocking against an immovable object, it bounces up into the air. Once entrained the particle is carried by the wind describing a parabolic path and strikes the ground with considerable force. When it strikes the ground it either bounces up again or cause other particles to bounce. The height attained by sandgrains depends on the velocity of wind, the nature of the surface and also on the velocity of the grains. This type of transportation which is carried out through a series of bounces is called *saltation*.

The minimum wind-velocity needed to start the saltation process is about 20 Km per hour. Particles carried by saltation causes more abrasion. Laboratory experiments have shown that a high speed grain in saltation can move by impact, a surface grain six times its own diametre or more than two hundred times its own weight.

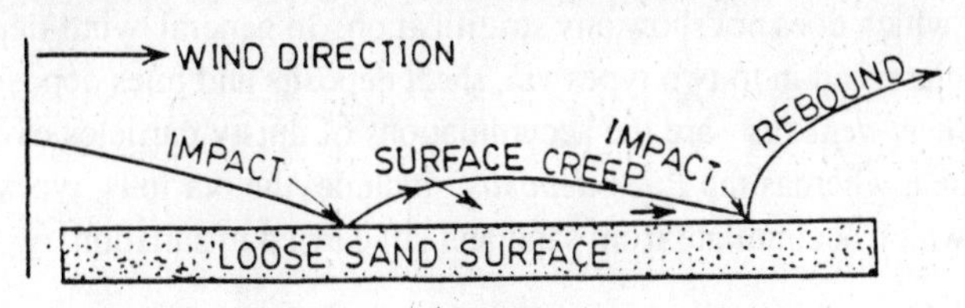

FIG.66.

Surface Creep

Particles too large and heavy to be lifted off the surface by the blowing wind may be gradually and erratically moved along the surface. During normal high winds pebbles may be rolled along the ground surface but in majority of the cases when wind blows away the finer supporting material from beneath the boulders, they start rolling forward. This method of transportation, through rolling and creeping, is also known as *traction*.

DEPOSITION

As we know, the power of the wind to transport the particles which are made available to it by the erosional processes largely depends on its speed. Whenever the velocity of wind is checked due to some or other reasons, a part or whole of the load starts getting deposited. Drifting material carried by the wind tends to be heaped up by the slightest obstacle and eventually leads to the formation of what is known as *aeolian-deposits*.

Accumulation of rock-particles etc. transported by the wind may begin due to various factors as

(i) Relief of the locality,

(ii) Prevailing wind-velocity, and

(iii) Amount and distribution of vegetation cover etc.

Wind is an excellent agent for sorting of materials according to their size, shape or weight. Pebbles and boulders cannot be carried away and are left back to form *Lag deposits,* while the clayey and silty fractions are carried farther than the heavier and larger ones and float for a considerable time in air and then settle to form what is known as *Loess,* which does not show any stratification. In general, wind-deposits are classified in to two types viz, sheet deposits and piles deposits.

Sheet-deposits are the accumulations of dutsty particles over a large area, whereas the *Piles-deposits* include the various types of dunes which accumulate from sand and silt carried in saltation.

Loess

It is a deposit of wind-blown silt and clay particles 70 to 85 percent of these particles are in the range of 0.007 to 0.003 mm in diameter. Loess has a high porosity and is unstratified, since it accumulates in such fine layers that they do not show up. Loess deposits are usually buff-coloured. Loess-deposits are quite conspicuous in Northern China. The loess-region of north-western China is larger than France, and in some parts of it the loess has accumulated to a depth of over 300 metres, and are light-yellow in colour.

Loess deposits in the Mississippi-Valley are called *adobe*, where it was probably derived, in the main, from glacial outwash.

Deposits of Sand

Sand deposits are found in a variety of forms which also very in shape and size. Accordingly there are small-scale as well as large-scale features usually of sand deposits. Small ripples and ridges found on sand surfaces are the small-scale features, whereas the sand hills and various types of sand-dunes etc. are the large-scale features.

(i) *Ripples* These are small, low asymmetrical ridges, about 2 to 5 cm high and stretching parallel to one another. Ripples are formed at right angles to the wind direction and stretch laterally for considerable distances. Invariably the coarser particles are found on the crests and trough since they are moved forward by creep and collect the leeward of the ripple where they remain because there is virtually no saltation to move them.

The ripples are similar to those formed in sediments by a water current.

Ridges are larger than ripples and have a maximum height of 15 cms, besides they consist of coarse sand granules and small pebbles. The ridges have more height because the wind can remove only the smaller particles and leaves the larger ones to accumulate.

(ii) *Sand hill* Mounds of sand whose surfaces are irregular are called *Sand hills.*

(iii) *Sand-dunes* It is an accumulation of sand shaped by the wind usually in the form of a round hillock or a ridge with a crest. As Arthur Holmes puts it *any mound or ridge of sand with a crest or definite summit is called a dune.* The forms, structures and distribution of dunes are controlled by the intensity and direction of the prevailing winds. Dunes are most likely to develop in areas with strong winds that generally blow in the same direction.

Most sand dunes are asymmetric in cross-section. In structure, generally, a dune has a long and gentle slope in the wind-ward side, a sharp crest and a short and steep- face towards the lee-side. The steep-face on the leeward side is developed because the sand blown over the crest falls into a wind-shadow and comes to rest at its natural angle of repose which is between 20° to 35° for dry sand. Once formed, a dune acts as a barrier itself and with further deposition gradually grows in size.

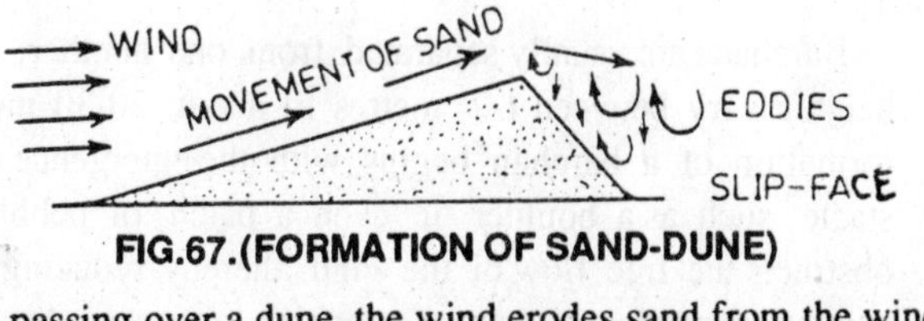

FIG.67.(FORMATION OF SAND-DUNE)

In passing over a dune, the wind erodes sand from the windward side and deposits the same sand on the slip face. In this process, the dune moves forward. In majority of the cases, a dune may move only 10 to 20 metres per year.

The formation of a sand-dune and its shape is controlled to a great extent by the following factors:

(a) Amount of sand supply,

(b) Wind-velocity,

(c) Constancy of wind direction, and

(d) Amount and distribution of vegetation cover.

The dunes which constantly change form under wind currents are know as *live-dunes* or *active dunes*, while the inactive dunes, which have become more or less stabilised by vegetation cover, are known as *fixed-dunes*.

The mineral composition of the sand grains in sand-dunes depends on both the character of the original sand source and the degree of chemical weathering in the region. Most dunes are composed largely of the mineral *quartz*. Besides, dunes are also found to contain grains of ferromagnesian minerals, gypsum and calcite grains.

Types of Dunes

A number of sand-dunes have been recognised, on the basis of their shapes, as follows:

(i) *Barchans* These are asymmetrical, crescent shaped dunes in which the points or wings of the crescent are directed down wind. These dunes are formed in strong winds and are situated perpendicularly to the main direction of the moving wind. They develop, generally, on a flat floor, with wind blowing from a constant direction and where the sand supply is limited.

Barchans are usually separated from one another, and their heights vary between 1-2 metres to about 20-30 metres. The formation of a barchan begins with the emergence of an obstacle, such as a boulder or even a patch of pebbles, which obstructs the free flow of the wind thereby reducing its speed and transporting power. Deposition occurs on the windward side of the obstacle and the sand then piles up until the dune attains a shape which offers the least resistance to the flow of the wind. Gradually wind is deflected around the sides of the growing dune, where its frictional contact with the sand slows down the wind to a little extent. Sand is deposited there in the form of long, pointed horns or wings, following the direction of the wind and thus giving the dune its characteristic crescent-shape. These dunes, generally, have a gentle windward slope and a much steeper leeward slope.

Brachans tend to arrange themselves in chains extending in the directin of the prevailing wind and they also gradually move forward. Speeds of their movement as great as 47 metres per year have been recorded in Peru.

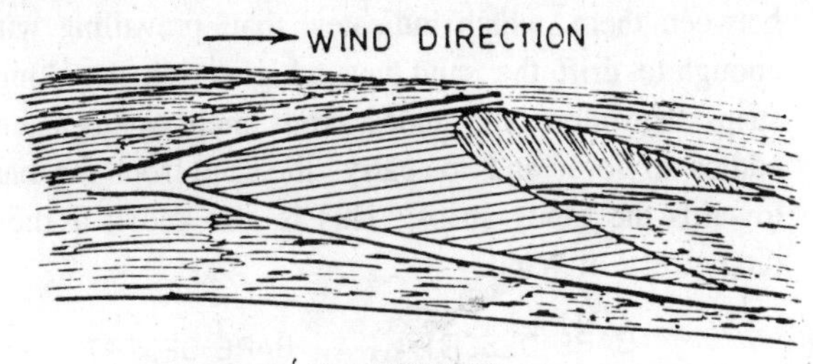

FIG.68. (BARCHAN)

(ii) *Longitudinal or Seif dunes* These are large symmetrical ridges of sand parallel to the wind direction. While some authors have distinguished seifs from longitudinal dunes, others use them as synonymous terms. According to E.W. Spencer (Basic Concepts of Physical Geology, Oxford & IBH Publishing Co. New Delhi, 1970) the seifs are dunes similar to the barchans except one wing is missing, which is caused by an occasional shift in wind direction but not in the direction from which sand is being supplied; whereas the longitudinal dunes form when sand is in short supply and the direction of the wind is constant.

According to Aurthur Holmes (Principles of Physical Geology. The English Language Book Society and Nelson 1975) *Linear ridges or longitudinal dunes (known as seifs in the Sahara), which commonly occur in long parallel ranges, each diversified by peak after peak in regular succession like the teeth of a monstrous saw.*

Even though there exists a divergence of opinion about the

term, majority of the authors accept seifs as longitudinal dunes. These dunes develop in areas where the sand supply is scanty and winds are strong from one direction. These dunes sometimes attain immense sizes, about 100 Kilometres in length and a hundred metres or more in height. These are crowned with low summits at regular intervals.

It seems likely that the longitudinal dunes or seifs are shaped by a prevailing wind from one direction, interrupted by occasional cross-winds.

Parallel series of longitudinal dunes often have bare desert between them, which indicates that prevailing winds strong enough to drift the sand have a higher velocity along the bare strips than along the dunes. Thus there occurs a tendency for eddies to form and to carry the sand from the bare patches towards the sandy strips. This is also known as the 'shepherding effect' of Wind.

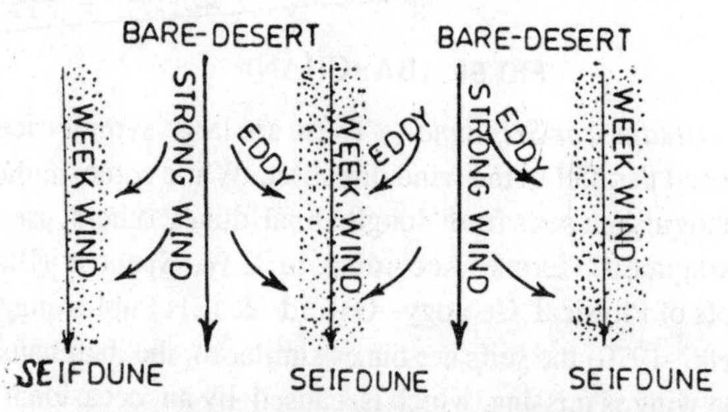

FIG.69. (SHEPHERDING EFFECT AND SEIFS)

(iii) *Transverse dunes* These are elongated dunes form at right angles to the prevailing wind. They develop where sand supply is abundant and there is no vegetation to interfere with the growth of dunes. These are relatively straight and take the form of wave-like ridges separated by trough-like furrows. These dunes may be very long but commonly have low-heights. They are of common occurrence along coasts and lake shores.

(iv) *Parabolic dunes* These dunes are parabolic in outline and they have their wings pointing towards the direction opposite to that of the blowing wind i.e. the curve of the dune crest is bowed convexly downwind, and the horns pointing upwind. They commonly form around blow-outs, particularly adjacent to beaches, where the sand supply is abundant and the wind direction is steady. Here the slip face on the downwind side is quite steep.

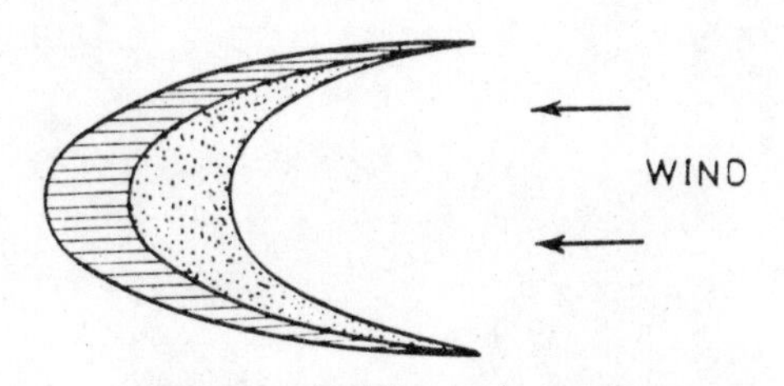

FIG.70. (PARABOLIC DUNE)

(v) *Whale-back dunes* These are very large longitudinal dunes with flat tops. They are also known as *sand levees.* Sometimes on the top of the whale-back dunes barchans or seifs may occur.

(vi) *Pyramidal or star-shaped dunes* These dunes are also known as *Rhourds.* They seldom stand alone and normally contain steep-radiating ridges culminating in one or several peaks. They appear to have been formed due to the interference of air-waves caused by wind reflecting from mountain barriers.

CONCLUSION

In spite of the fact that wind is an active agent of the nature, constantly interacting with the Earth's surface and that it has immense power, it has been noticed that as an agent of erosion and transportation, the action of wind is very much limited in comparison to that of running water and moving ice. Wind is only effective as an agent of erosion and transportation where the surface is made up of loose dry particles and where there is little plant cover to protect it from the wind.

23

GEOLOGICAL WORK OF UNDERGROUND WATER

All the water occurring below the earth's surface is termed variously as *underground water*, *subsurface water*, or simply as *ground water*. Underground water is universally present and constantly moistens the rocks with in its range. Part of the water resulting from atmospheric precipitation percolates through the soil into the rock strata forming underground water. This percolation is possible since the rocks forming the earth's surface, do always have some openings present in them and through them water can conveniently sink downwards. At very great depths, however, the enormous pressure of the overlying rocks effectively reduces the number and extent of these openings, thereby fixing a lower limit below which underground water can not occur.

The subsurface occurrence of groundwater may be divided into two great zones, known as the 1. Zone of saturation and 2. Zone of aeration. Both of these zones are separated from each other by the Water-Table.

1. *Zone of Saturation* It is that zone of rocks where all rocks are saturated with ground water. As such, all the openings and pores are fully filled with water in this zone. The top surface of it is called the Water-Table or Phreatic surface.

2. *Zone of Aeration* Above the water table and below the surface of the earth, the pore spaces and other openings are only partially filled up with the water that percolates through them. This is the zone of aeration and in this zone, destructive chemical action takes place. It is the zone that Van Hise has denominated "the belt of weathering", in which the oxygen of the atmosphere assisted by moisture, carbonic acid, the organic acids and locally by sulphuric acid, acts on the rock and produce the manifold products of weathering.

The water occurring in this zone is known as *Vadose Water or Suspended Water.*

The occurrences of ground water is chiefly controlled by the following factors:—

(a) *Cimate* In arid climate ground water generally occurs at great depths, whereas in humid regions it occurs at shallow depths in flat areas, at less shallow depth in hilly areas and at near the surface in the valleys. The water table rises in wetweather but sinks in dry weather.

(b) *Topography* When the surface is almost flat, the water table remains parallel to the surface. In an undulating humid region the water table approximately follows the undulations of the surface but is less accentuated, rising under hills and flattening out under valleys, often intersecting the valley floor so that the occupying river is fed.

(c) *Properties of rocks* As we know, the materials underground, however are heterogeneous and their hydrologic characteristics are varied. These variations control the occurrences of ground water. Porosity and permeability are two of the most important properties of rocks that control the occurrence of ground water to a major extent.

Porosity

It refers to the percentage of interstitial spaces in a given volume of rock. Generally the open spaces connect with each other. Compaction and the filling of pore spaces by a cementing medium decreases the porosity of a rock. The porosity varies in different rocks on the basis of the following factors:—

(i) *Sorting and size of the grains* Coarse, rounded and uniformly graded sediments lead to high porosity compared with fine, angular and unassorted sediments. Thus, it seems likely that a boulder conglomerate would be more porous than a sandstone.

(ii) *Shape and packing of the grains* Rounded particles can differ in porosity according to the way they are packed. Minimum porosity results when the spheres are offset and maximum porosity occurs when each spherical particle is directly above the centre of the other.

As such, hard and compact rocks like the igneous and metamorphic ones have very little open space in them and the porosity is usually less than one percent, whereas loose sand or gravels may have porosity to the extent of 25-45 per cent.

Permeability

It is the ability of a rock to transmit water through it. Permeability is more commonly defined as the relative ease with which water moves through the interstices of the rock. Therefore, for movement of water the pores must be interconnected. Some sand dunes and unconsolidated gravels have a high porosity and are also permeable since the large pore spaces are interconnected. But it should be remembered that a rock having high porosity may not necessarily be highly permeable. For example, muds may have high porosity, but the pore spaces are so small and poorly connected that water can not move between them. A rock may be porous without being permeable but cannot be permeable without having pores.

According to the degree of permeability, rocks are subdivided into three groups as:—

(i) *Permeable* Gravels, sands, fissured sandstones, limestones etc.

(ii) *Semipermeable* Sandy loams, loess etc.

(iii) *Impervious* Clays, shales, non-fissured massive, crystalline and cemented sedimentary rocks.

Ground water occuring in permeable geologic formations are known as aquifers. Rocks, though porous but not permeable enough to furnish an appreciable supply of water for a well or spring are known as aquicludes. The rock which is impervious to water is known as aquifuge.

SOURCE OF GROUND WATER

Ground water has gotten into the earth from one of the following sources:

1. Meteoric water.
2. Condensational water.
3. Connate water.
4. Juvenile water.
5. Mixed source water.

Meteoric Water

It includes waters formed by infiltration of atmospheric precipitation like rain, sleet, snow, hail etc. as well as by the infiltration of water of rivers and lakes. As we know, falling on the earth's surface, atmospheric precipitation is distributed in a number of ways—a part of the water enters the soil by infiltration, a part is returned to the atmosphere through evaporation and another part flows over the ground surface as runoff to lower levels. Water of such origin apparently constitutes the bulk of ground water, which is evident from the fluctuation of water level in the wells with the changing of seasons.

Condensational Water

This water is mainly the source of replenishment of ground waters particularly in deserts and semideserts, where precipitation is scanty and there is rapid evaporation. In such regions, there is ground water at certain depth below the surface. This is believed to be due to the following process:

The moist air at the surface of the earth is always warmer than the air of the soil, particularly in summer. Accordingly, there exists a difference in the pressure between the water vapour in the atmosphere and in the soil. Because of the pressure gradient, water vapour from the atmosphere penetrates the rocks and are converted to water through condensation when the temperature falls. This may lead to the accumulation of a certain amount of water in rocks in arid and desert regions.

Connate Water

This is also known as fossil water and includes water entrapped in sediments at the time of their deposition on lake or sea bottom. They are classified in to two types as syngenetic and epigenetic connate water. The syngenetic connate water was trapped in the sediments containing it, whereas the epigenetic connate water are those which entered from the basins into the rocks that had formed earlier. Connate water often occurs in rock units with oil.

Juvenile Water

It is also known as magmatic water as it is associated with the magmatic activities with in the crust. With the cooling of magma, its gaseous contents and water vapour etc. separate out from it. The water vapour then gets condensed into superheated water and move upwards from a region of high temperatures and pressures to that of low temperature and pressure. This is also called virgin water.

Mixed Source Water

It is quite natural to expect that along their complex migration routes the aforesaid waters get mixed up and thus constitute ground water of a mixed type.

Geological Action

The geological action of the ground water involves processes like erosion, transportation and deposition.

Erosion

As we know, ground water circulates slowly under the influence of gravity, in general, following the slopes of the water-table. The rate of water movement is very slow and vary considerably from one place to another according to the permeability of the rocks.

Mechanical processes of erosion are absolutely insignificant in case of underground water. Ground water brings about erosion mainly through chemical processes i.e. by the solution action, which is particularly effective in regions of soluble rocks like limestone, dolomite, gypsum and anhydrite, rocksalt etc.

The dissolving capacity of ground water depends on its content of carbon dioxide and free oxygen as it enters the zone of aeration and increases with a rise in temperature and pressure. When underground water circulates through limestone and other soluble rocks it tends to dissolve the material along the joints and bedding planes producing passages and chambers of all dimensions. Underground water is particularly effective in the solution of limestones and produces a distinctive landscape together with underground caves. The topography thus developed is known as *Karst-topography*, after an area of that name in northeastern Italy and northwestern Yugoslovia, where it was first investigated in detail. Karst topographic features develop both above and under the ground surface.

IMPORTANT EROSIONAL FEATURES

(a) *Lapies* The leaching action of the ground water as it passes through the limestone region, produces a highly rugged topography. Where the limestone is exposed at the surface, water running across the surface gives rise to straight rounded grooves with sharper ridges in between. These are called *rillenk-arren.* The ground water may enlarge the joints of the lime-stone into a conjugate pattern of clefts and ridges. This surface is called lapis-surface or limestone-pavement. The clefts in such a pavement are called *grikes* and the ridges *clints.*

(b) *Sink* It is a large solution cavity, may be several metres in diametre, open to the sky. These are also known as *dolines, sink-holes* or *swallow-holes.* Sometime the sink-holes become so numerous that the sides begin to touch one another. Surface drainage becomes limited to short sinking creeks, those that disappear in to the ground. Along such streams there are small holes where water swirls into small openings leading into caverns.

(c) *Caverns* These are interconnected subterranean cavities in limestone, formed by the solution action of ground water. These cavities are always having roofs intact. Caverns always vary considerably in size within wide limits and are sometimes of exceptionally large dimensions and are commonly inter-linked. The horizontal linking passages are known as *galleries* and the inclined or vertical ones as *shafts.*

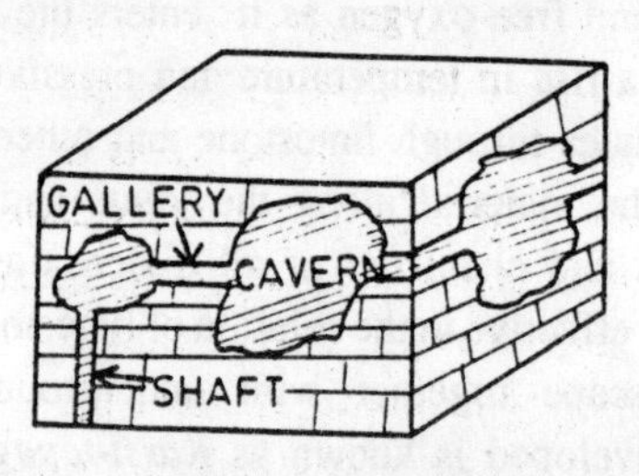

FIG. 71. CAVERNS IN LIMESTONE REGIONS

(d) *Solution Valleys* Sometimes the roof of a cavern may collapse, enlarging it in an upward direction. With continued solutions, the collapse may reach the surface and a hollow, usually elongated and narrow, may be produced and form a solution valley. These valleys are normally developed on limestones and are also known as *dry-valleys.* They resemble the channels formed by running water on the surface. Many such valleys are also called *blind-valleys* as the water from their streams is lost to subsur face channels.

In case of partial failure of the roof on top of a cavern, there occurs a natural bridge of limestone over-arching the solution valley. These natural bridges are, thus, the remnants of the roof of a cavern.

(e) *Polje* These are large depressions, occurring mainly due to the roof collapse over great Karst chambers. They are characterised by their extensive size, flat bottom and the shape of a closed basin with steep sides. They are often filled with water forming polje lakes. Small residual hills found on the floor of poljes are called *Hums* or *pepino-hills*.

(f) *Stylolite* It is an irregular suture like boundary developed at the junction of two consecutive soluble rocks, where the less soluble portions of the consecutive beds projects into each other.

TRANSPORTATION

Transportation by underground water takes place in solution. Sometimes they are carried to sea or lakes through percolation and sometimes they are added to the stream water.

DEPOSITION

The dissolved materials, travelling in solution are subjected to precipitation and consequent deposition due to the following factors:—

(i) Evaporation.

(ii) Loss of Co_2

(iii) Chemical reaction.

(iv) Change in physico-chemical condition (i.e. temperature and pressure).

IMPORTANT DEPOSITIONAL FEATURES

1. *Stalactites and Stalagmites* Particularly in caverns lying above the water table, water may be seen dripping from the straight line of a crack in the ceiling of the cave. These droplets lose some of its dissolved gas by evaporation and deposit small granules of calcium carbonate at the point of the evaporated drop. In due course, with subsequent deposition of calcium carbonate (by the process as stated above) it grows downwards as an icicle-like pendant. These deposits hanging from the roof taper towards the floor and are called *Stalactites.*

The water that drops from the end of the stalactite falls to the floor of the cave immediately beneath it and with continued evaporation and deposition domeshaped or conical deposits of calcium carbonate start growing upwards. Such deposits are called Stalagmites.

Stalactites and stalagmites together constitute what is known as *drip-stone.*

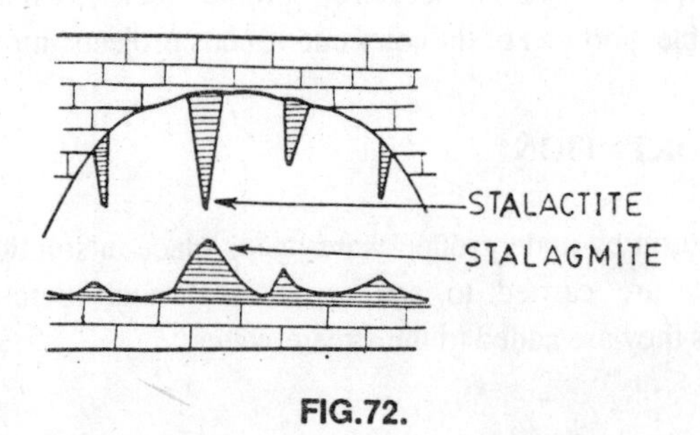

FIG.72.

2. *Sinter and Travertine* Ground water emerging on the surface as springs or seepages form loose superficial deposits of silica or calcium carbonate. The silica deposits thus formed at or near the exit of such springs are known as siliceous sinters whereas those formed of calcium carbonate are called *travertine* or *calc-sinter.*

3. *Geode* Deposits resulting from the ground water often forms incrustations on the walls of the cave or fissure, below the water-table. Successive layers of amorphous silica thus formed differ in their colour and texture, and are known as agate. But, when crystallisation is

perfect, well developed crystals of quartz etc. are found tapering towards the centre of the cavity as the teeth of a comb. Such deposits are known as Geode-which are partially or completely filled cavities.

4. *Concretion* These are spherical, nodular, discoid or lens-like forms and are due to concentrated solution effects of groundwater on rock substances which are rather difficultly soluble. These are deposits formed around some nucleus which differ in chemical and mineral composition from the enclosing substance. In clays and shales the concretions are mainly calcareous and in sandstones they may be ferruginous or calcareous.

5. *Replacement deposits* By the process of replacement, ground water dissolves matter present in any particular substance while depositing an equal volume of material it contains. Thus it is some sort of a substitution on a volume-for-volume basis and the new material preserves the most minute texture of the one replaced. Shells of fossils are replaced in this way; even petrified wood is also a result of such replacement.

24

GEOLOGICAL WORK OF GLACIERS

As Davis suggested a glacial topography is a climatic accident that happens to normal cycle of erosion, i.e., climate gets very cold and the river freezes. Instead of rivers of water there are rivers of ice, called 'Glaciers' which operate as the main geomorphic agent.

Longwell and Flint have defined a glacier as follows:—

A glacier is a body of ice, consisting mainly of recrystallized snow, flowing on a land surface.

[Introduction to Physical Geology Published by John Wiley & Sons, New York and London-1961]

In simple terms a glacier may be defined as a large natural accumulation of ice with a downward or outward movement from the snow-field under the influence of gravity. They move down the pre-existing valleys or radiate out in great lobes.

At present it is estimated that approximately 10 percent of the earth's land surface and 7 per cent of the oceans are covered by glacier ice. The largest extent was attained during the Pleistocene or Great Ice Ages.

Formation of Glaciers

Glaciers originate in snow fields. Glaciers are formed through the accumulation of snow and its subsequent transformation. In areas where annual snow fall exceeds the amount of loss of snow due to evaporation and melting, snow starts accumulating. Snow accumulates when the air temperature is too low for it to melt. As we know, low temperature prevails in regions of high latitudes and at the mountain summits where there is not enough warmth in the summer time to melt the accumulated snow. Thus the snow mass grows from year to year.

Ice forms only where the snow cover is thick and continuous and is not melted during a summer thaw. A number of factors play important roles in the process of changing snow into ice. These factors are:-

1. A low average annual temperature.
2. Continuous and abundant precipitation falling as snow.
3. The altitude in the mountainous regions and the latitude of a locality which effects the formation of glaciers to a major extent; since the climatic zones of the earth are related to the degree of latitudes.
4. Existence of a suitable landscape and relief forms.

With the falling of temperature below 0°C, some atmospheric moisture is precipitated in the form of hexagonal ice crystals commonly known as snow-flakes. Newly fallen snow is highly porous with a specific gravity of 0.05. Successive snow falls bring about the compaction of the lower layers and the increasing pressure causes a slight lowering of the melting point of snow. Besides, in the summer the snow at the surface begins to thaw, the snowflakes melt and at night during the recurrence of frost they refreeze. A part of the melt water seeps deeper into the loose snow through the interstices and melts the snow-crystals further down. When the melt water refreezes around the snow crystals they assume the shape of grains. Thus ice granules are formed. As the snow becomes granular its porosity diminishes and specific gravity increases. This process is known as *regelation*. The granular masses of ice are called ne' ve' in French and *firn* in German language. Firn is a dull white, impermeable and structureless form of ice.

This process, as described above, is repeated with each new

snowfall. With further addition of snow and under the influence of pressure the firn becomes more compact and the separated crystalline aggregates merge into frozen masses forming what is known as *glacier-ice*. The glacier-ice is transparent and has a bluish tint. It has a specific gravity of 0.8 and is impermeable. The glacier-ice constitutes the main body of glaciers. Thus Arthur Homes has defined glaciers as follows:

Glaciers are masses of ice which, under the influence of gravity, flow out from the snow fields where they originate.

[Principles of Physical Geology, Published by the English Language Book Society & Nelson 1975]

As we know, temperature drops by 0.5° to 0.6°C per each 100 metre rise in altitude; besides the temperature also decreases with an increase in the latitude i.e. from the equator to the poles. In any region, the lowest limit of perpetual snow is known as the permanent snow line or commonly as *Snow Line*. Above the snow-line are the snow fields that persist through out the summer season. The height of the snow line varies with the latitude of the place and are modified to some extent by the local climatic conditions and topographic pattern. In the polar regions it lies at altitudes close to sea-level (Antarctica); in Norway and Alaska it is at an altitude of 1500 metres, in the Himalayas at 5100 metres in Assam-region and approximately 6000 metres in Kashmir region. It is important to note that permanent snow fields occur in almost all the continents except Australia.

An upper limit of the snowline has also been imagined on the basis of the fact that the moisture content of the atmosphere diminishes with the rise of altitude and that there may be conditions unfavourable for snow to form at some particular height. This gives an impression that the peaks of the mountains would be free from snow even after their tops reach that height. The term 'Hionosphere' is used for the region between the Permanent Snow Line and the imaginary Upper Snow Line.

Glacier Movement

A characteristic feature of glaciers is their ability to flow. When sufficient ice gets piled up, it exerts substantial pressure on the ice in the lower layers and it acquires plastic properties which enable the ice-mass to move outward or downhill and thus an active glacier comes into being.

When there is thick accumulation of ice on a slope the glacier under the influence of gravity begin to flow slowly down a valley until it reaches a point where the rate of melting exactly balances the increment of ice. A glacier may extend far below the snow-line. When the flow of ice exceeds the amount lost by melting the glacier advances, and if the melting is in excess, the termination of the glacier recedes up the valley and the process is called the retreat of a glacier.

On level surface the movement of ice is not ordinarily noticed. In such cases, the thickness of the snow in the snow-fields increase indefinitely and the flow is along the radii from the centre to the periphery. As we know, the accumulation of snow is more in the central part of the glacier and there is a decrease in the thickness of ice near the marginal parts. Accordingly the pressure on the lower-layers of ice in the central parts is more where they acquire plasticity and start moving along the radii towards the periphery. The melting of glacier becomes more towards the peripheral region and the glacier thins out closer to the margin. Thus differential pressure on the glacier plays an important role in its movement on a level surface.

Glacier motion is very slow and its velocity (i.e. the rate of movement) varies from a few centimetres per day to a few metres per day. The rate of movement depends on a number of factors, such as:

(i) Thickness of the glacier;

(ii) Gradient of the slope which it covers;

(iii) Temperature of the ice;

(iv) Rate of evaporation and melting;

(v) The intensity of retarding friction along the slope, etc.

Apart from the above, it is also important to know that the movement of the glaciers is more in the central parts than at the sides. At the sides frictional drag retards the rate of movement and therefore the lateral parts move slowly in comparison to the middle parts. Some mountain glaciers move very rapidly and are called *Surging Glaciers*.

The rate of movement of the Himalayan glaciers varies from 2 to 4 metres per day. The glaciers of the Alps move at the rate of

0.1-0.4 metres per day. Some of the glaciers of Greenland move at a rate of about 20 metres a day.

Types of Glacier

On the basis of their stage of development, size, shape and the relationship between the supply and flow areas, three types of glaciers have been distinguished. These are as:—

1. Mountain or Valley glacier;
2. Piedmont glacier;
3. Continental ice-sheets;

1. *Valley Glaciers* These are also known as Mountain or Alpine-type of glaciers. They are confined to the pre-existing valleys in mountain areas and are fed by snow-fields which lie further up, above the snow-line. The ice flows down the mountain valleys with steep slopes to heights determined by the rate of supply from above and the rate of melting as it reaches warmer levels. They frequently descend below the regional snow-line.

The valley glaciers are characterised by a distinctly expressed supply area i.e. the snow-field where the snow is converted into firn and then into ice and a drainage area i.e. the area over which the glacier-ice moves and flows. While the supply area is situated above the snow-line the drainage area is made up of mountain valleys situated below the snow-line.

According to their characteristic features several kinds of valley glaciers have been distinguished, such as:—

(i) Simple glaciers, which are isolated glaciers consisting of single flow without any tributaries.

(ii) Complex or Polysynthetic glaciers, consist of a number of coalescing glaciers, the pattern of which resembles that of a river with tributaries.

Apart from the above, the types of glaciers included in the category of valley glaciers are as follows:

Cirque Glaciers

These glaciers originate in deep arm-chair shaped hollows situated at the valley heads. They are often at the snow-line and with hardly any flow.

Transection Glaciers

These are valley glaciers which have become so thick that they spill over dividing ridges and join with other glaciers in adjoining valley. These are also known as *avlanches* or Hanging Glaciers.

In the case of valley glaciers the topography controls the motion and the movement is in one direction only, down the valley slope mainly due to gravity and the glaciers move several metres a day.

Valley glaciers are common in the young fold mountain areas such as the Alps, Himalayas, Tienshan, Pamirs and Caucasus. The Hubbard glacier in Alaska is the longest valley glacier in the world with a length of about 130 kms. Most of the Himalayan glaciers are small. The Gangotri glacier is about 24 kms. in length and the Siachan glacier is about 72 kms long.

2. *Peidmont Glacier* These are also known as Intermediate type of glaciers. They are intermediate in form as well as origin between the valley glacier and the continental ice-sheets. Sometimes in colder climates valley glaciers may extend over a low-land and spreads out horizontally. Several glaciers thus unite at the base of a mountain range forming an extensive and comparatively thick sheet of ice covering the low-lying ground. Such an ice-sheet is called a piedmont glacier. The Malaspina glacier of Alaska is the best known example of piedmont glaciers.

These glaciers are much larger in dimension than the valley glaciers. Their rate of movement is quite slow.

In contrast to the formation of ice-sheet at the foot-hill region of a mountain range the intermediate type of glaciers also include the formation of ice-caps on the flattened surfaces of the summits of ancient mountains covering them for hundreds of square kilometres. These are also called plateau glaciers. Like continental ice-sheets they lie as a continuous mass covering a huge area on the plateau and in moving from the centre to the margins, these glaciers emerge through river-valleys and resemble mountain glaciers. Thus plateau glaciers combine the characteristics of both continental ice-sheets and valley glaciers. Such glaciers are common in Scandinavia (Norway) and are therefore sometimes referred to as Scandinavian-glaciers.

3. *Continental Glaciers or Ice-Sheets* These are the largest forms of accumulation of ice and they cover vast areas of the landmass including even the cliffs of mountains. At present they occur mostly in Antarctica and Greenland. These glaciers are of enormous size and immensely thick. The thickness of the ice-sheet may reach even thousands of metres, as such all irregularities of relief are hidden by it. The topography has little or no control over the movement of such ice-sheets. The surface of the ice-sheets has a plain-convex shape which resembles a shield. Their shape is not controlled by the bottom relief. Unlike the valley glaciers, they do not have distinctly separate supply and drainage area.

The movement of ice is radial. The movement of the ice is in many directions from points of high pressure within the ice-sheet towards the margin. The movement is very slow which takes place at the bottom, while the top of the ice-sheet remains almost stationary.

At the margins, the thickness of theice-sheet diminshes and mountain peaks and individual cliffs project through the ice, which are called *nunataks* by the Eskimoes.

Glaciers of the continental type are formed in polar regions and are located almost at sea-level. At present, the Antarctica ice-sheet is the biggest continental type of glacier. In Greenland the ice-sheet covers almost the whole of the continent. The Antarctic ice-sheet covers 13 million square kilometres.

The complete ice-sheet does not reach the sea. The ice which enters the sea tends to float upon the water and its marginal part is buoyed up. It is thus easily broken by the waves giving rise to separated ice-mass known as *Icebergs*. These icebergs are large pieces of ice floating on the sea.

An important glacial feature of Antarctica is the presence of huge plates of floating glacial ice, known as ice-shelves. As the ice-sheets are not confined by valley walls the ice over-flows the coasts and presses out to sea as colossal floating ice-barriers (e.g. the Great Ross Barrier) from which the great tabular icebergs of the Antarctica seas break off.

Surface Features of Glaciers

The surfaces of glaciers are usually rough and uneven because of the presence of gaping fissures known as *Crevasses* which may be open and visible, but are often masked by snow. The surface part of the glacier is brittle. Brittleness of the surface part makes it crack as it is subjected to tension, whereas the ice beneath behaves like a plastic substance and moves by slow flowage. Crevasses are seldom more than 30 metres deep and 7 metres wide. The flow of ice at depth prevents the formation of crevasses at depths of 30 metres or so.

Movement through mountain valleys and major irregularities of the earth's surface, gives rise to differential movement within the mass of glacier which results in the development of crevasses. However, the formation of crevasses is conditioned by a variety of factors as follows:

(i) Relief of the subglacial bed;

(ii) Variations in the cross-section of the valley through which the glacier moves,

(iii) Thickness of ice,

(iv) Rates of flow of glacial ice i.e. the differential movement at the middle and marginal part of the glacier, etc.

On the basis of the mode of formation and nature of the cracks, crevasses are classified into three main types as :

— Transverse crevasses.

— Longitudinal crevasses.

— Marginal crevasses.

Transverse Crevasses

When there is a slight change in the gradient of the valley or there is a marked steepening of its slope the glacier is subjected to a considerable tension which is relieved by the development of a series of cracks, transverse to the direction of flow, at the bends. Since the glaciers move more rapidly in the middle than at the sides, these crevasses become curved with the convex side facing downward.

When the change of gradient is more pronounced the glacier is broken into a huge jagged mass of ice pinnacles known as *Seracs*. These are similar to waterfalls in a river. Accordingly they are also known as ice-falls. An abrupt steepening of the slope forms what is known as *rock-step*.

Longitudinal Crevasses

When there is a sudden widening of the valley, the glacier expands sideways and assumes its shape. Such spreading out may develop cracks which are more or less parallel to the length of the glacier i.e. parallel to the direction of flow.

Marginal Crevasses

These crevasses are formed due to the differential rate of movement of glacier at the middle and the marginal parts (i.e. the valley sides). While the middle part of the glacier moves more rapidly, the marginal parts move quite slowly. Accordingly cracks are developed along the valley sides which are oblique to the course of the glacier and are pointing to up-hill direction. These crevasses are also known as lateral-crevasses.

A wide and very deep crevasse that opens near the top of the firn field of a cirque where the head of a glacier is pulled away is known as the Bergschrund (in German). Such crevasses usually open in summer.

As the glacier creeps down the slope to the foot-hill region where the gradient is gentle the cracks close up and the crevasses disappear.

The presence of debris on the surface of the glacier tends to cause rapid melting and thus sometimes melt a hole in the ice which are known as *dust-wells*. The dust-wells often unite forming a depression of the shape of a bath-tub, commonly called *bagnoire*. The melt water gathering into streams mostly fall into crevasses and by their melting and pot-hole action deep cauldrons are formed in the glacial ice, which are known as *glacier mills or moulins*. The water escapes to the front of the glacier through a tunnel.

GEOLOGICAL ACTION OF GLACIERS

The geological action of glaciers comprises erosion, transportation and deposition which together constitute what is known as glaciation. The geological action of glacier is mainly due to its flow. Since the piedmont glaciers are intermediate in nature and characters between the valley glaciers on the one hand and continental ice-sheets on the other, the geological action of glaciers can best be studied separately for valley glaciers and continental ice-sheets.

Geological Action of Valley Glaciers

The most significant action is carried out by valley glaciers. As it has already been explained, the valley glaciers flow down the pre-existing stream valleys and reshape the valley in specific ways.

Erosion

The erosive action of a glacier takes place due to (i) plucking, (ii) rasping and (iii) avalanching. The erosive action of glacier is more pronounced particularly when the thickness of the ice is great, the pressure on the subglacial floor is high and the glacier ice is heavily charged with rock fragments.

Plucking

This is the process in which the moving ice lifts out blocks of bedrock loosened by the freezing and thawing of water in fractures beneath the ice. Water due to rain or melting often seeps down along the sides of the ice mass filling up the cracks, fissures and porespaces within the country-rocks along the edges and at the head of the glacier. When the temperature drops, this water freezes within those openings and exerts enormous pressure on the country rocks due to expansion in volume, breaking them up. The broken blocks are frozen in suspension in the ice and are carried away along with the ice. Thus, plucking involves two processes-quarrying and frost-wedging. While frost-wedging causes a shattering of the country- rocks, the quarrying process lifts out the shattered blocks, of rock. The plucking process particularly affects the downstream side of outcrops of well-jointed rocks.

Rasping

The process is also known as *abrasion* or *corrasion.* Glaciers normally carry considerable quantities of rock fragments in their basal sections. These rock fragments are dragged over rock surfaces, and their sharp points and edges cause characteristic scratches, gougings and grooves in the underlying hard bedrocks. Since the rock fragments are dragged under great pressure over bedrock, they themselves are scratched and worn down into peculiar facetted stones. The sharp points and edges of the rock fragments are gradually blunted by friction. Thus, some glacial scratches and grooves which may start by being narrow and deep, gradually become broader, and shallower and finally fade out. This serves as an indication of the direction of ice movement in a given place; the movement of ice is from the deep and narrow end of the groove towards its broader and shallower end.

Bare rock surfaces are scraped and scoured due to abrasion. If the under surface of the glacier is studded with rock particles consisting of silt or sandgrains the rock beneath will be polished; if they are gravel or boulders the rock will be scratched, or striated (if it is softer than the fragments). While the rock-studded bottom of a glacier functions as an effective file or rasp and polishes, scratches and abrades the surfaces over which it moves, the front edge of the glaciers function like a bulldozer pushing and scraping the ground in front of it and is more effective in soft and semiconsolidated sediments.

Avlanching

This is a process of mass-wasting. When the valley sides are scraped and the rock debris which are broken off are carried away by the glacier ice, there results a great deal of under-cutting, of the valley side. This leads to mass-wasting, bringing huge amounts of debris onto the top surface of the glacier.

Erosional Features Produced by Valley Glaciers

The most diagnostic features of glacial erosion occur in areas of high relief and precipitation. Most of the evidence of glacial erosion available is in the shape of landforms which are presumed to have arisen largely by erosion. Some of the major features produced by glacial erosion are as follows:

1. *The Cirque* This is a French term for amphitheater shaped basins commonly located at the head of a glaciated valley. These are steep-sided semicircular depressions often with their floors overdeepened to rock basins. In different countries they are variously termed, such as *Kar* in German, *Cwm* in Welsh, *Corrie* in Scotch, *Botn* and *Kjedel* in Scandinavia.

These bowl-shaped depressions are excavated mainly by frost action. A cirque begins to form, beneath a snowbank or snow-field just above the snow-line, by snow accumulation in small erosion rills or other chance depressions in the slope of a mountain. The rock surface beneath and around the snow bank is gradually broken up and deepened by freeze-thaw and mass-wasting. while the rock surface is broken up by the freezing of water, the smaller rock particles are carried away downslope by meltwater during thaws. Such shattering by freeze-thaw action is thought to be responsible for the overall enlargement of the depression. This process of quarrying of rocks by frost action is called *nivation*.

The steep head wall (or backwall) seems to be produced largely by the shattering of the rocks in freeze-thaw alternations. The steep headwall is relatively free of talus at its base. It may be a kilometre or so in height. With more active excavation of the floor the depression grows larger and ice starts accumulating to greater thickness. When enough ice gets accumulated, it flows downslope as a glacier. Thus cirques are the main sources of supply of valley-glaciers.

After the glacier melts, a small lake known as *Tarn* usually occupies the depression. A cirque basin usually terminates at a bedrock riser called its threshold. A small glacier in a large cirque is indicative of the waning stages of glaciation.

2. *Aretes* As already described, the cirques grow steadily larger due to frost shattering and plucking. The headward erosion gradually consumes the preglacial uplands. In this process, two adjacent cirques along the opposite slopes of a mountain may begin to coalesce which results in a typical jagged knifelike ridge, known as an *arete* or *comb*. They are often referred to as *razor-edged- ridges*, serrate-ridges, or *sawtoothed-ridges*.

3. *Horn* This is a sharp, pyramidal peak produced due to growth and enlargement of three or more cirques together by headward erosion. The up land being consumed by cirque-erosion from several sides is reduced to a number of aretes radiating from the central summit. In due course the aretes themselves are worn back and the central mass remains as a pyramidal peak.

4. *Col* It is a depression formed along the arete at a place where the headwalls of two opposed cirques intersect each other.

5. *Glacial Troughs* Valley glaciers reshape their valleys by widening, deepening and straightening them. Most glacial valleys were originally stream-cut valleys. The typical V-shaped cross section of the stream-cut valleys are transformed to a wide, deep, flat-floored and steep-sided U-shaped valley, known as *Glacial-trough.* The long profiles of the glacial valleys is irregular and ungraded which exhibits a series of step-like forms and deep basins cut from the bed-rock. The step-like forms are known as *glacial-step* or *glacial- stairway*. Each step consists of three components known as a riser, which marks the down valley end of each step; a riegel which is a sort of rock bar at the top of a riser; and a tread, which is relatively a flat surface of a step. The steps and basins are believed to develop because of the differential resistance of

the rocks composing valley-floor to glacial erosion. Riegels develop where the bed rocks are massive, unjointed and highly resistant to erosion, whereas basins are formed in highly fractured and jointed zones of the bed rock which offers least resistance.

6. *Hanging Valleys* Tributary glaciers also carve out U-shaped troughs. Tributary glaciers generally contain less ice than the main valley glaciers and accordingly in the latter case the floor is more deepened than the tributary glaciers meeting it from the sides. After a period of prolonged glaciation the tributary valley appears to hang above the floor of the main-valley occupied by the larger glacier. Such tributary valleys are called hanging-valleys. The junctions of such valleys are the sites of waterfalls.

7. *Truncated Spurs* As already explained, the rock-studded bottom of the glacier can abrade the valleys both laterally and vertically in a more effective manner. Besides, glaciers are less easily deflected by obstacles than rivers and the flow of ice is streamline. The sharp and acute bends which occur along the course of stream-valleys are straightened up by glacial abrasion and in the process it cuts projecting spurs producing blunt tringular facets, known as *truncated-spurs* or *facetted-spurs.*

8. *Glacial boulders* The rock fragments entrapped in the glacial ice, get abraded, rounded and their surface polished and striated during the course of the glacier movment. Such rounded blocks are called glacial-boulders.

9. *Glacial Scars* These are small-scale erosional features produced by the abrasive action of the glacier. Rock debris entrapped by the glacier during its movement are dragged under great pressure over the bed rock. The sharp points and edges of the rock debris produce characteristic scratches, gougings and grooves in the solid rock. The process of glacial abrasion also produces polished and facetted surfaces. These erosional features are collectively referred to as glacial scars.

10. *Roches Moutonnees* These are also known as *Sheep-back rocks,*

formed due to glacial abrasion. When a little hill of rock or small elevation is encountered on the way of the glacier, it is not usually worn away completely. The side facing the direction of glacier movement becomes gentle, smooth and striated, while the opposite side remains rough, rugged and steep. The gentle sloping up-hill side is also called the *stoss-side* and the steep down-hill side is called the *lee-side*. While abrasion is pronounced on the stoss-side, plucking is prominent on the lee-side.

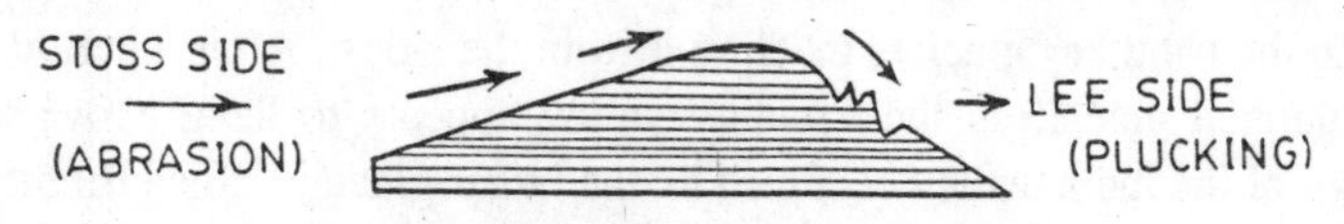

FIG. 73. ROCHES-MOUNTONNES

These forms are somewhat elongated and longitudinal asymmetrical. From a distance a group of such features often look like sheep lying down or resemble the curls on a lawyer's wig. Rocks thus shaped are called *roches-mountonnees*.

11. *Fiords* These are deep glacial troughs which have been eroded below sea-level or are submerged to a depth below sea-level (even though they are formed at higher levels). As a result the U-shaped valley may continue right out into the sea, which are subsequently occupied by sea-water itself, and become inland stretches or arms of sea. Within fiords, the glaciers come in contact with the sea-water where blocks of ice break off from the glacier and float on the sea as icebergs. This process of wastage of glacier is known as *Calving*.

Transportation

Glaciers are effective agents for transporting enormous quantities of rock debris. The load consists of rock debris ranging in size from fine particles to huge boulders, which are distributed more or less evenly through out the whole mass of the glacier. A large part of the load of valley glaciers is contributed by :

— erosion of rocks by the glacier;

— weathering and mass-wasting processes which operate quite effectively in mountain slopes, above the level of the glacier ice:

— plucking of the unconsolidated materials over which the glacier moves.

Much of the load of the valley glaciers is carried on the glacier-surface and on its sides. The load is carried frozen within the ice, however, larger and heavier rock-fragments sink down through the ice and accumulate at the bottom of the glacier, which is responsible for the rasping and abrasive action of the glacier. The load carried by the glaciers is known as *Glacial-drift*, which is unsorted and unstratified.

Deposition by Valley-Glaciers

The transporting power of a glacier diminishes when the ice begins to melt and the glacier slows down resulting in the deposition of the material carried by the glacier itself. Deposition takes place mainly in the downstream parts of glaciers, where the load is dropped forming huge accumulations of varying shapes and characters. In glacial deposits big boulders and finest rock materials are accumulated at the same place. Glacially transported and deposited materials together constitute what is known as *Glacial-drift*. This may be stratified or unstratified. When dropped directly by the glacier, there results a heap of glacial deposits consisting of an assemblage of rock debris of varying dimensions; whereas when reworked subsequently by the meltwater streams issuing from the decaying and retreating glacier there results sorted and stratified deposits. Accordingly, glacial deposits are classified into two main types as:

(a) Unstratified and unsorted deposits.

(b) Stratified or Glaciofluvial deposits.

On the basis of their mode of formation glacial deposits may exhibit a variety of forms as described below:

DEPOSITIONAL—FEATURES PRODUCED BY VALLEY-GLACIERS

Unstratified Deposits

The unstratified, unsorted debris dropped more or less in a random fashion by glaciers form deposits known as *till*. Till makes up a group of topographic features called *Moraines*. The distribution of moraines marks the extent of now-vanished glacier.

The essential character of a till is that it is formed of a mass of broken rock fragments mostly angular or subangular and of many different sizes and compositions. The term *Boulder-clay* is also used to describe such deposits. Consolidated moraines are called *tillites.*

Moraines are localized deposits of glacial debris formed either on the body of the existing glacier or at various places along the glaciated valley of an extinct glacier. Accordingly, two different types of moraines are recognized

1. Moving Moraines.
2. Stationary Moraines.

1. *Moving Moraines* These moraines travel with the glacial ice. Depending on their position on the body of a mountain or valley glacier various types of moving moraines have been distinguished, as indicated below.

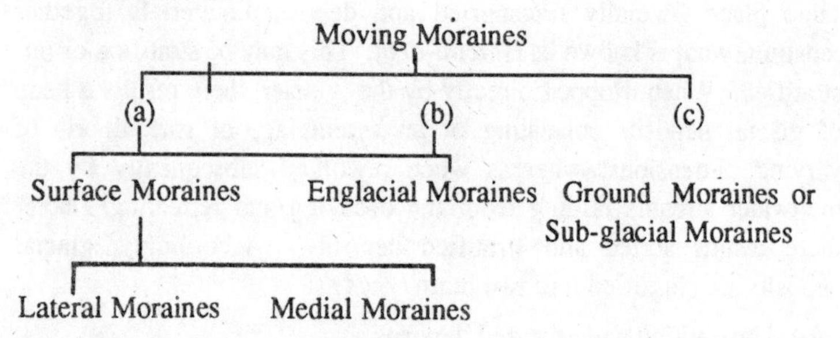

(a) *Surface Moraines* Weathering and mass-wasting processes which operate efficiently in mountain slopes above the level of glacier contribute to the rock debris carried by the glacier on its surface. Such accumulation of rock debris on the surface of the valley glacier is known as *surface-moraine.* Surface moraines are in turn divided into two types as

(i) Lateral Moraines.

(ii) Medial Moraines.

Lateral moraines are low-ridges of rock debris formed along the margins of the glaciated valley. The rock debris are scoured from the valley sides above the glacier due to glacial erosion and also by weathering, snowslides, avalanches and other types of mass-wasting. Besides, the rock debris are also brought by streams along lateral gorges. These rock debris are carried on the surface of the glacier along the margins where they have been dumped from the valley sides, forming what is known as *lateral moraines* or *marginal moraines.*

Medial moraines are formed when two glaciers merge through the confluence of valleys. When two glaciers join, their lateral moraines are dragged along and their coalescence give rise to longitudinal ridges in the middle part of the glacier. Thus the lateral moraines of tributary glaciers subsequently become the medial moraines of the main glacier. Because of their central position they are quickly removed away and are therefore transitory in existence.

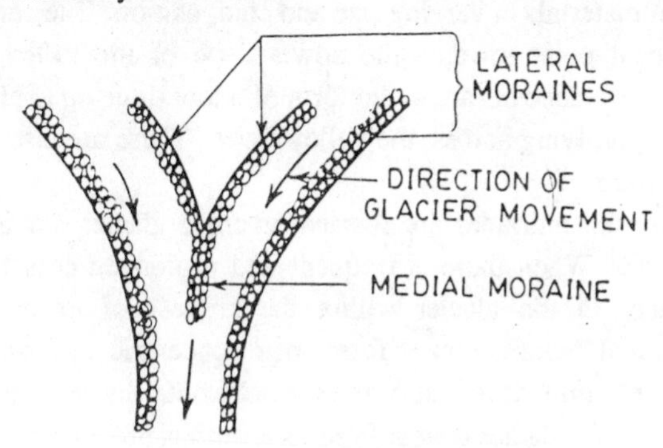

FIG. 74. (LATERAL & MEDIAL MORAINES)

(b) *Englacial Moraines* The rock debris lying on the surface of a glacier may fall into crevasses during the movement of the glacier and are enclosed in the body of the glacier. This debris enclosed in the glacier's body is known as *Englacial moraine.* Sometimes there is an accumulation of rock debris, due to weathering processes, in the neve field where these are burried under the newly fallen snow and later on engulfed in the moving ice they reach the drainage area.

(c) *Ground Moraines* These are also known as *Sub-glacial Moraines* or *bottom moraines.* As we know, when a glacier moves over the valley-floor rock-fragments are plucked out, dragged along and incorporated in the basal layers of ice. As the the glacier retreats, the materials carried by the glacier are dropped at random upon the valley floor forming what is known as *Ground-moraine*. These deposits are quite irregular in form and are typical examples of glacial till. These deposits are thin and uneven.

2. *Stationary Moraines* These are also known as deposited moraines. In such cases the rock debris is deposited after melting of the glacier. These moraines include the terminal and recessional moraines.

Terminal moraines are formed at the terminus of glaciers, where the ice front remains stationary for long period of time. Various fragmental material are deposited at the terminal fringes of glaciers when the ice melts away as a crescent-shaped areas of tumultuous heaps consisting of materials of varying size and composition. The convexity of the crescent point towards the down slope of the valley where they occur. These deposits are in the form of a low ridge up to about 30 metres in height, lying across the valley floor. These are also known as *End Moraines*.

Recessional moraines are formed where a glacier retreats in a halting manner. When there is frequent and prolonged cessations in the movement of the glacier within the process of its recession, several terminal moraines may form in a concentric fashion which indicate the intermittent retreat of the glacial front. Successive pauses in the position of a glacier-retreat from its terminus produce successive moraines known as *recessional moraines*. These moraines may be pushed ahead to a new position if the glacier readvances.

Apart from the moraines the unstratified glacial deposits also include features like erratic blocks.

Erratic-Blocks

The term erratic has been derived from the Latin *erraticus*, meaning wandering. The large boulders, plucked by the glacier from the bedrocks and carried over a great distance are dropped and come to rest on the land surface where the country rocks are of distinctly different types. The boulders are quite huge in dimensions. Such glacier-transported blocks are called erratic-blocks. Some of the boulders rest in precarious positions on abraded surfaces. These are known as *perched blocks*, which also provide evidence of the direction of glacier movement. Pocking and logging stones are the erratic-blocks delicately balanced upon the glaciated bedrock. A group of erratics spread out fanwise is a *boulder train*.

Stratified or-Glaciofluvial Deposits

These are the deposits of streams formed by the melting of glaciers. These deposits are sorted and stratified by the action of water from melting ice. While the melt water deposits the coarse material near the end of the glacier, the finer material is carried further away. The water of the streams flowing out of glaciers are often milky-coloured due to the presence of rock flour which is a product of glacial abrasion. The milky-coloured water is, therefore, known as *Glacial-milk*.

The important depositional features of glaciofluvial origin include outwash plain, valley trains, eskers, kames and kame terraces.

Outwash Plains

Beyond the terminus of most mountain glaciers the gradient are commonly less steep and the valleys are wider which help in the deposition of rock material carried by streams fed by glacier melt-water. The deposits are formed as alluvial fans, which coalesce into a gently sloping extensive plain known as 'Outwash plain.' These are composed of stratified sand, gravel and shingle. While the coarser fragments like gravels and shingles are usually deposited near the edge of the terminal moraines; the accumulation of sandy deposits are formed farther away. The outwash material may extend for hundreds of kilometres down major valleys.

Valley Trains

Outwash sand and gravel occuring on valley floors form what is known as valley-train. They appear as terraces beginning from the terminal moraines and extending down the valley.

Eskers

An esker is a sinuous ridge extending in the direction of the movement of glacier. These are steep-walled ridges of assorted and stratified gravel and sand. Their length ranges from a few hundred metres to several tens of kilometres and are from 3 to 30 metres or more in height. These are a few tens of metres in width. Some eskers are more or less straight, while others are shaped more like a river-valley. Even though there is a divergence of opinion regarding the origin of eskers, mostly these are believed to have been formed from the deposits laid down by the water flowing in the superglacial, englacial or subglacial stream channels. The glacial streams, as we know, transport the load of the glacier which comprises sand, gravel and pebbles and deposit them in the glacial channels in the manner very much similar to that of rivers. When the glacier finally melts away, the deposits of the glaciofluvial streams are left standing as a ridge on the glacier floor.

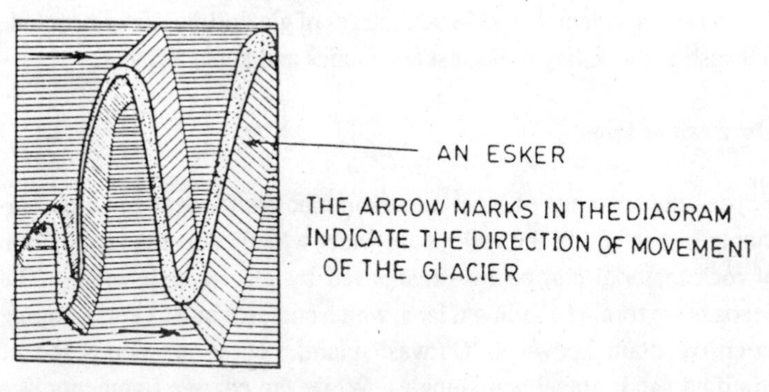

FIG.73. ESKER

Some authors believe that with the emergence of the sub-glacial streams into the frontal region of the glacier its velocity decresses resulting in the deposition of material carried by it at the margin of the glacier in the form of a narrow delta. More and more new deltas are formed with the continuous retreat of the glacier. Such deltas may coalesce to form a continuous ridge known as esker. But this process of formation fails to explain the rise of an esker on to an elevation and therefore this opinion does not receive much acceptance. Eskers are also known as Osser or oss.

Kames

These are small rounded hills, composed of stratified sand, gravel, pebbles and fine clays with an average height of 10 to 12 metres. They are often found near terminal moraines. The presence of fine clays indicates that these are formed in stagnant water and are also associated with immobile ice.

Due to intensive thawing there forms numerous depressions and basins on the surface of such ice, which are filled with water and are thus turned, into small lakes. The surficial meltwaters carry sediments from the top surface of the glacier into these depressions. As the ice melts, the material that formerly filled depressions on top of the glacier gradually sinks down. With the complete melting of the glacier, these deposits take the form of small, irregular and rounded hills, known as *kames*.

Kame Terraces

These are ice-contact features in which the deposits are laid down against an ice surface. It is formed by the fillings of depressions between the glacier and the sides of the trough. Sometimes these are confused with lateral moraines but the composition of the material distinguish one from the other.

Apart from the above important depositional features, features like crevasse infillings, kettles etc. are associated with glacial deposition. Surficial melt water fills the crevasses with the material carried by them which form short and straight ridges known as crevasse-fillings. This is a special variety of kame.

Kettles are depressions developed on the surface of glacial deposits. These are either circular or elongated in shape and vary in size between wide limits. These hollows are formed by melting of blocks of ice that might have been enclosed or burried within the glacial drift during the process of deposition. Such depressions mark the former location of the ice block. Sometime a few of such depressions may form lakes or swamps. These features are produced by stagnant glacier ice. The outwash plains are often pitted with such depressions.

Glacio-Lacustrine Features

Lakes are quite common in a glacial topography and are filled up of the sediments carried by the streams issuing from the glacier. When the ice disappears or the lake is drained away, layers of fine clay and silt accumulated in the lake bottom get exposed.These deposits are called glacio-lacustrine sediments which are stratified and form flat lacustrine plains.

GEOLOGICAL ACTION OF CONTINENTAL ICE-SHEETS

Erosion

As we know, the continental ice-sheets consist of large masses of glacial ice covering an extensive area where the surface features are buried under them. Despite their comparatively sluggish movement, the ice-sheets are also capable of eroding the solid bedrock. The erosion of the bed-rock is accomplished mainly by plucking and rasping (abrasion). Mass-wasting process, which plays a significant role in case of Valley glacier is not at all responsible for supplying rock debris to the ice-sheets. This is mainly due to the fact that the ice-sheet starts moving only when it becomes sufficiently thick and at this stage, its upper surface is usually above the mountain peaks. It is only for this reason that the mass-wasting process cannot operate effectively in the erosion by continental ice-sheets. The plucking and rasping process contribute much to the load carried by the ice-sheets. They are carried frozen into the bottom of the ice-sheet or are pushed ahead by it.

Erosional Features Produced by Continental Ice-Sheets

Continental ice-sheets tend to remove the irregularities of the topography which are encountered in their path and inhibit their movement. Rock debris entrapped by the glacier or frozen into the bottom of the glacier enables the glacier to abrade the underlying hard bedrocks and produce features like striations, grooves and polished surfaces on the bed rock over which the ice move and on the rock debris carried at the bottom of the ice. *Crescentic gouges* are formed on the bed rock where rock fragments moving in the ice encounter obstructions in their path and eventually chips off pieces of the bed rock. The *chatter marks* which are in the form of a series of curved cracks along the direction of glacier movement, are produced when sharp pointed rock fragments dragged over the surface of the bedrock by the moving ice. It is to be noted that the curvature of the crescentic gouges are opposite to the curvatures of the chatter marks.

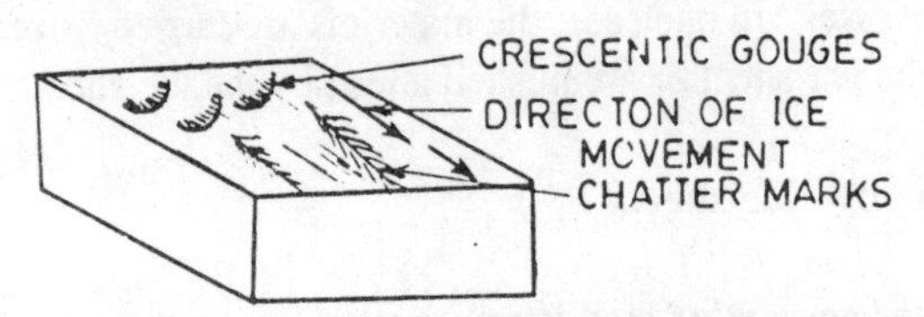

FIG.74. CRESCENTIC GOUGES AND CHATTER MARKS

The erosional features like striations, grooves, polished surfaces, crescentic gouges and chatter marks, as described in the preceeding paragraph, constitute the small scale features produced due to erosion by the continental ice-sheets.

The large-scale erosional features produced by the continental ice-sheets include the followings

(i) Roches moutonnees (These features have already been described under valley-glacier)

(ii) *Crag and tail* When very hard rocks like igneous bodies or volcanic plugs are encountered on the path of the moving ice-sheet, they stand as pillars and offer resistance to the flow of the

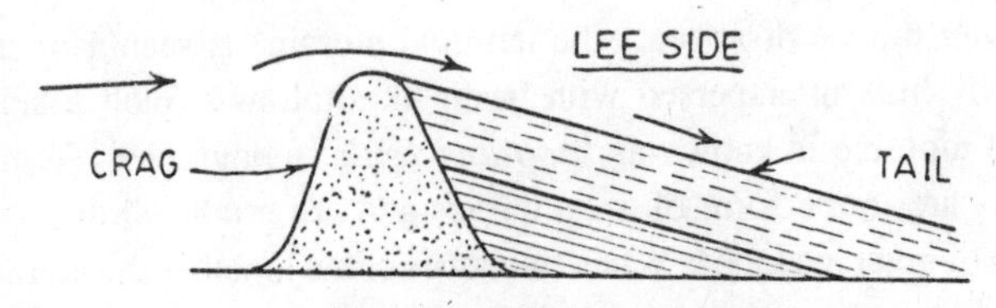

FIG.75. CRAG AND TAIL

ice and retard its erosive action on the rocks behind the obstructions. This produces a feature in which the obstructing block facing the moving ice is known as *crag* followed by a sloping lee side made up of softer rocks called *tail*.

Apart from the above features, ice-sheets make enormous excavations on the weak bed rocks. In places, the ice-flow is accentuated by the presence of a valley and behave much as a valley-glacier and produce many of the features of valley glaciation.

Transportation by Ice-sheets

Transportation of materials by the ice-sheet is accomplished by two distinct processes. In one case, the materials are carried frozen into the bottom of the ice-sheet or are pushed ahead by the ice-sheet.

Deposition by Ice-sheets

An enormous amount of load is transported by the moving ice. When the ice begins to melt and the glacier slows down, the load carried by it gets deposited. All the rock materials tend to accumulate at the terminus of the ice-sheet. As in the case of valley glaciers the deposits formed by the ice-sheets are also of two types, such as - Unstratifed deposits (till) and Stratified deposits.

Unstratified deposits which contain a hetrogeneous mixture of rock debris ranging in size from fine particles to huge boulders without any size assortment are called till, as already discussed under valley glacier. The ice-sheets produce only three types of moraines as - Terminal moraines, Recessional moraines and Ground moraines. It is to be noted that the lateral and medial moraines which mainly occur in valley glaciers are absent in the continental ice-sheets.

After the ice disappears the terminal moraine is seen as a chain of knobby hills interspersed with basin like hollows. Such a belt of terminal moraine is known as *knob and kettle* topography. Terminal moraines are in the form of great curves and are produced due to the advance of great ice lobes. When two lobes come together the moraines get fused into a single moraine between two ice lobes. Such a deposit is known as an *interlobate moraine*. The terminal, recessional and ground moraines have already been discussed under the depositional features produced by valley-glaciers.

Drumlins

These are moraines, often composed of till. These are relatively low, smoothly rounded, oval hills with a steeper face (stoss side) upstream. The long axis of these features are parallel to the direction of movement of the glacier. Drumlins may have the length ranging from a few hundred metres to 1 or 2 Kilometres and in height they may range from 10 to 60 metres. They are usually two to three times as long as they are wide. Each drumlin resembles the inverted bowl of a spoon.

Drumlins are invariably found in a zone behind the terminal moraine. They always tend to occur in groups which may number in the hundreds; single drumlins are rare. The topography thus produced is called a *basket of egg-topography*. Drumlins are formed beneath the moving ice when the clay-rich debris is abundant near the base of an ice-sheet. As a result, any knob-like obstruction in the path of the ice gets plastered above and around with dense clay and boulders, which are then moulded into a low streamlined hill called a *drumlin*. Thus a bedrock core is often observed at the middle of the drumlin. The blunt end of a drumlin faces the direction from which the ice flowed and thus serves as an indicator of direction of ice movement.

Drumlins are mainly associated with Continental ice-sheets and show features opposite to those of *roches mountonnees*.

FIG.76. DRUMLIN

Glociofluvial Deposits

These are sorted and stratified deposits formed by the melt water issuing from the glacier. These deposits include the depositional features like outwash plains, eskers, kames and kettles, kame terraces, erratics etc. which bear a close similarity with those of valley glaciation.

Glaciolacustrine Deposits

These deposits include features like lacustrine-plains, varves etc. The lacustrine plains are of similar characteristics with those of valley glaciers.

Varves are thinly laminated clay and silt deposits formed in the calm and quiet environment of a lake, by the streams of melt-water issuing from the ice.

EVIDENCES OF GLACIATION

Glacial erosion is indicted by the following features:—

(i) Striated, polished and grooved bedrock surfaces.

(ii) Presence of unweathered bedrock material.

(iii) U-shaped valleys.

(iv) Hanging Valleys.

(iv) Cirque.

(v) Unsorted and facetted rock fragements.

(vi) Presence of erosional features like roches mountonnees, crags and tails etc.

(vii) Presence of depositional features like moraines, drumlins, eskers, kames, kettles etc.

CAUSES OF GLACIATION

A number of hypothesis have been advanced to explain the causes of glaciation of the earth's surface, but all of them are based on some speculations. The periods of glaciation are very unusual and the causes giving rise to periodic glaciation fail to explain a number of observations, a few of which are as follows:

— It has been established that there has been a number of periods of glaciation separated by warm, moist interglacial periods. But there is no supporting evidence to explain that the glacial periods occurred at regular intervals.

— The quick growth and decay of ice-sheets.

— Formation of glaciers in mountainous regions or landmasses at higher altitudes.

The causes for glaciation are as follows:

(i) Variations in the eccentricity of the earth's orbit and the inclination of its axis to the plane of the ellipse take place periodically which affects the amount of radiation received from the sun at any one place at a particular time. During the period of greater eccentricity the distance between the sun and the earth also changes. The earth's position is at a maximum distance from the sun once every 92000 years, when the conditions will favour a temperature drop and consequent glaciation. This indicates that glaciation should take place at regular intervals, but records of glacial climate do not furnish any evidence of glaciation at regular intervals. This hypothesis was advanced by Milutin Milankovitch, a Yugoslovian astronomer

(ii) Variations in the intensity of solar radiation intercepted by the earth may bring about a general glaciation on the earth's surface due to increased snowfall and cause the accumulation of snow and ice in highland areas and mountains until the ice sheets are formed and begin to advance. With the increase of radiation the ice melts and retreats causing intrglacial periods in the earth's history. But, it is also possible that increased solar radiation can cause glaciation and whether increased or decreased solar radiation causes glaciation still remains unanswered.

Apart from the above lines of explantion for the variation in solar radiation, F. Nolke (1937) suggested that due to the passage of the sun through a nebula there is absorption of a part of the sunrays resulting in the decrease of solar radiation which ultimately cause periodic spells of cold climate and glaciation. It has been suggested that variations in the density of the nebula is responsible for the alternation of glacial and interglacial periods. This theory was challenged because of the fact that nebulas are so rarified that they are incapable of reducing the solar radiation enough to have any substantial effect on change in the earth's climate as a whole.

(iii) According to Swedish scientist S. Arrhenius, changes in the composition of the atmosphere, particularly the carbon dioxide content, may cause a drop in the temperature which in turn causes glaciation. Carbon dioxide of the atmosphere freely passes the sun rays to the earth's surface but absorbs the reflected heat rays (i.e. the infra-red rays). Therefore, an increase in the carbondioxide content would lead to a rise of the temperature while a decrease in the carbon dioxide content would bring about a fall of temperature. It has been estimated that if the carbon dioxide content of the atmosphere is reduced by half, the earth's average surface temperature will go down by 4°C. This may bring about the growth of ice-sheets to cause glaciation. It is thought that abundant vegetation helps in the depletion of carbon dioxide content of the atmosphere bringing about a fall of temperature.

(iv) Variation in the amount of dust in the atmosphere influences the earth's climate in a substantial way. Increased amount of dust in the atmosphere absorbs more solar energy and reduces the solar heat reaching the earth's surface. This may cause a lowering of the air temperature, which creates conditions favour able for glaciation.

(v) Shifting in the positions of the continents with respect to the poles have been accounted for the glacial climate of a particular area. On the basis of the palaeo-magnetic studies, it has been established that the north and south poles have not always been in the position they now occupy in relation to the continents. Thus, shifting of the continents causes the landmasses to move to favourable geographic positions for the growth of ice-sheets and consequent glaciation.

(vi) Elevation of continental masses by tectonic movements changes the relief of the landmasses. Increase in elevation of large parts of the continents, mainly due to widespread orogenic movements, give rise to mountain systems. With the increase in the height of the crust, the temperature falls as a rule and conditions favourable for the accumulation of ice are created in the mountain ridges situated above the snowline. Besides, mountains intercept much precipitation by the orographic mechanism. Thus an upheaved landmass is capable of changing the climate of a particular area and may cause increased snowfall over favourable highland areas of the continents which could cause glaciation.

Past Glaciations in India

1. The geological history of India reveals the records of the earliest glaciation from the Kaldurga-conglomerate, in South India, of Dharwar age.
2. Vindhyan glaciation, which took place about 600 million years back, has left evidences mainly in Madhya Pradesh.
3. Gondwana glaciation occurred about 300 million years back.
4. Pleistocene glaciation which includes five glacial and four inter glacial periods occurred in India between 1 to 3 million years ago.

25

GEOLOGICAL WORK OF OCEANS

It is well known that about 71% of the surface of the earth is covered by the oceans and seas. The oceans and seas cover an area of about 361 million square kilometre out of 510 million square kilometre of the surface of the entire globe. About 1.4 billion cubic kilometres of water is concentrated in oceans and seas. The greatest known depth in the ocean is 11022 metres at the Mariana Trench in the Pacific. Land is concentrated mainly in the northern hemisphere and the water bodies in the southern hemisphere. Nearly 61 per cent of the area in the former and 81 per cent in the latter are covered by water. The four recognised oceans in the world are - the Pacific, the Atlantic, the Indian and the Arctic ocean. The Pacific Ocean covers about 49% of the earth surface, the Atlantic Ocean—26%, Indian ocean—21% and Arctic ocean - 4% of the world ocean.

The geological activity of seas and oceans, like other geological agents, comprises the processes of erosion, transportation and deposition, which depend on a large number of factors such as :

(i) Relief of the floor.

(ii) Chemical composition of the sea water.

(iii) Temperature, pressure and density of sea water.

(iv) Gas regime of seas and oceans.

(v) Movement of sea water.

vi) Work of sea organisms etc.

(i) *Relief of the floor* It has been established that the floor of the oceans exhibits an uneven topography with prominent elevation and depressions. On the basis of available bathymetric maps, the ocean is divided into definite regions as indicated below:

(a) *Continental shelf* The ocean floor gradually slopes downwards away from the shore. The shallow-water zone adjoining the land, with average depth down to 200 metres constitutes the continental shelf. It varies in width from a few kilometres to several hundred kilometres. The continental shelves cover about 7.6 per cent of the total area of the oceans and 18 per cent of the land. About 20% of the world production of oil and gas comes from them.

(b)*Continental slope* From the edge of the continental shelf, the sea floor commonly descends to the ocean basin, with an average gradient of 3.5° to 7.5° and is known as continental slope. Its depth ranges between 200—2500 metres and covers about 15% of the total area of the ocean. It has an average width of 16 to 32 Kilometres

(c)*Continental rise* It extends from the bottom of the continental slope to the floor of the ocean basins. The rise has a slope of 1° to 6°. Its width varies from a few kilometres to a few hundred kilometres. The material of the rise has been derived from the shelf and slope.

(d) *Ocean floor* It begins at a depth of 2000 metres and goes down to 6000 metres. It covers 76 per cent of the total area of the oceans and has a very gentle gradient, being measured in minutes. It contains a number of distinctive topographic units such as abyssal plains, seamounts and guyots, mid-ocean canyons, and hills and rises that project somewhat above the general level of the ocean basins. Apart from these features, the most significant feature of the ocean floor is the occurrence of mid-oceanic ridges and deep-oceanic trenches.

(ii) *Chemical composition of sea water* The oceanic water contains a large number of dissolved salts and have almost a uniform composition. These salts result in the property of salinity. The average salinity of sea water is 35 parts per thousand i.e. one litre of sea water contains 35 grams of various dissolved salt. But the value is small where large rivers meet the sea and the value is higher within the zone of hot and dry climate. In the Mediterranean Sea, for example, the sea level is lowered by evaporation and the salinity as well as the density of the water increases.

Sea water of normal salinity contains mostly chlorides which aggregate above 88% followed by sulphates more than 10% and small amounts of carbonates and other compounds. Sodium chloride constitute the bulk of the dissolved salts in sea water, followed by Magnesium-chloride, Magnesium sulphate, Calcium sulphate, Potassium-sulphate. Apart from these salts, there are also elements like iodine, fluorine, zinc, lead, phosphorous etc. in sea water.

Salinity determines features like compressibility, thermal expansion, temperature, density, absorption of insolation, evaporation, humidity etc. It also affects the movements of the ocean waters.

(iii) *Temperature, pressure and density of sea water* The temperature of the oceanic waters plays a significant role in the move ment of large masses of oceanic waters and distribution of organisms at various depths. The temperature of the oceans is not uniform. The temperature of the water on the surface of oceans and seas is determined by the climatic conditions. In tropical zones it is usually higher than in polar regions. Besides the temperature also varies with the depth of the sea water.

There are two main processes of the heating of oceanic waters, viz. absorption of radiation from the sun and convection; whereas cooling is caused by back-radiation of heat from the sea surface, convection and evaporation. The interplay of heating and cooling results in the characteristics of temperature.

The pressure in the oceans and seas varies vertically and in creases with depth by 1 atmosphere for each 10 metres of the water column. It is highest in the oceanic trenches (between 800-1000 atmosphere). At great pressure the dissolving capacity of sea water increases.

The density of sea water varies with narrow limits between 1.0275 and 1.0220, due mainly to variation of temperature and salinity. It is highest in higher latitudes and lowest in the tropical areas.

(iv) *The gas regime* The sea water contains mostly dissolved oxygen and carbon dioxide. Sea water derives oxygen from the air and also through photosynthesis by marine plants. Similarly the content of the carbon dioxide is mainly due to the atmosphere, river waters, the life activity of marine animals and volcanic eruptions. It has been seen experimentally that at a temperature of 0°C, the sea water can absorb about 50 cubic centimetre of carbon dioxide and 8 cubic centimetre of oxygen. The content of oxygen and carbon dioxide are of much significance in the processes of marine sedimentation and dissolution of chemical compounds.

(v) *Movement of sea water* The sea is a mobile mass of saline water and the movements of sea water are of great geological importance as they determine the intensity of destruction caused by the oceans and seas on the shore and the floor and also the distribution and differentiation of the sedimentary materials that enter the seas and oceans. The waters of oceans and seas are subjected to the action of wind, the attraction of the sun and the moon, and to changes of temperature, salinity, density etc. All these factors give rise to three main types of movements. Such as Waves, Currents and Tides.

Waves

Waves are generated mainly by the wind blowing over the surface of the ocean and sea water. The friction of wind moving over the water surface causes the water particles to move along circular or near-circular orbits in a vertical plane parallel to the direction of wind. There is almost no forward motion. Thus energy is transferred from the atmosphere to the water surface by a rather complex mechanism involving both the friction of the moving air and the direct wind pressure.

The ocean waves are osciallatory waves (or transversewaves) as they cause an oscillatory wave motion. The waves consist of alternating crests and troughs. Wave length is horizontal distance from crest to crest or trough to trough. Wave height is the vertical distance between trough and crest. Wave period is the time taken by two consecutive crests to pass any reference point.

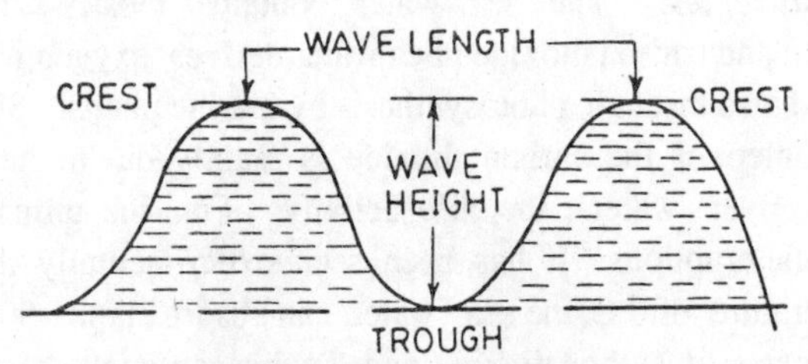

FIG.77. WAVE ELEMENTS

Wave velocity is the ratio between the Wave length and the Wave period.

The waves of oscillation are characteristic of deep water. As waves move into shallow water, they are slowed by the friction with the sea-floor and thus the wavelengths become shorter while the wave height increases, the paths become elliptical and the wave steepens. Since the the front of the wave is in shallower water than its rear part, there is an increase of the steepness of its frontal slope and the wave becomes highly unstable. At this stage, the wave is transformed into a breaker, which then collapses forward in a curling, frothing zone of surf. As the wave breaks, its water suddenly becomes turbulent. The turbulent water mass then moves up the beach as the *swash* or uprush. Thus, both water and wave energy move forward against the shore and the wave is called a Wave of *translation*. This energy causes erosion and transports material along the shore. The return flow which sweeps the sand and gravel sea ward is called *backwash.*

Currents

In the currents, there is an actual movement of the water over great distances, which may be caused by various factors, such as—the differences in temperature, salinity, action of steady and periodic winds etc.

Tides

Tides are periodic movements of the ocean waters due to the gravitational attraction of the sun and moon on the earth. Twice a day, about every 12 hours 26 minutes, the sea level rises and it also falls twice a day. When the sea rises to its highest level, it is known as 'high tide' and similarly when it falls to the lowest level, it is called 'low tide'.

There are two tides of a special occurrence viz. (a) the spring tide and (b) the neaptide. The spring tides occur twice every month at new moon and full moon, whereas in the first and third quarter the attraction of the sun and moom tends to balance each other and small tides, which are termed 'neap tides', occur.

The currents caused by high tides in the littoral zone are quite strong and can carry quite big fragments of rocks to the shore or along it, eroding the bottom.

(vi) *Work of sea organisms* Seas and oceans are inhabited by a large variety of animals and plants. Their development and distribution depends on the depth of the sea, its temperature, salinity, pressure, penetration of light and the dynamics of the sea water etc. Marine organisms are divided into three major groups such as benthos, plankton and nekton.

The benthos group includes organisms both mobile and sessile which inhabit the bottom of the sea. The plankton group includes organisms which are passively floating by the waves and currents. Unicellular organisms (animals) like foraminifers and radiolarians, and, diatoms (plants) belong to this group. The nekton group includes all actively swimming animals which comprises all the sea vertebrates and invertebrate molluscs.

These marine animals are important in producing biogenic sediments.

The above factors together play significant roles in the erosion, transportation and deposition by the oceans and seas.

MARINE EROSION

Seas and oceans does immunse erosion of the rocks constituting the coast and the bottom through the action of moving water. Erosion by the sea is accomplished mainly by the following processes :

(a) hydraulic action,

(b) abrasion,

(c) attrition, and

(d) corrosion.

Hydraulic Action

As it has already been explained, when a wave breaks, both water and wave energy move forward against the shore. The force of the water itself as it is thrown forward in the breaking waves bring about a remarkable wear and tear of the rock exposed to them. The striking force of waves breaking on the shore exerts sufficient hydraulic pressure that can shatter rock masses into fragments. Besides, air in joints and cracks is suddenly compressed by the impact of the wave and with its recession, the compressed air expands with explosive force pushing out large blocks of rocks.

The effect of hydraulic action becomes more conspicuous during intensive storms, when the waves hit the shore which can be as great as 38 tonnes per square metre which is sufficient enough to bring about erosion along the coast. Besides, the force of water alone is enough to be effective in dislodging fractured blocks.

The following factors affect the hydraulic action to a great extent.

(i) Strength of the rocks along the shore.

(ii) Structural features of the rock.

(iii) Configuration of the coast-line.

(iv) Force of the water

Abrasion

The destructive force of water is increased manifold when the water contains various rock fragments. Fragments of rocks picked up by the water are hurled against the shore as the wave breaks. The impact of large rock fragments thrown by large waves is capable of breaking up the most resistant rocks like granites and massive sedimentary ones, while the loosely consolidated rocks exposed along the shore are rapidly worn down by the impact of sands, gravels and pebbles by the waves.

Besides, the rock particles due to their back and forth movement over the bed rocks give rise to continuous rubbing and grinding (which has been likened to the effect of saw acting at a cliff-foot) wears down the bedrock lying along the shore. This process of wear and tear is known as abrasion. Abrasive action is faster and conspicuous where the rocks are soft and much jointed and slow where they are hard and strong.

Attrition

The rock fragments during their to and for movement along with the prevalent waves and currents, collide with one another and are themselves worn down, as a result the fragements are reduced in size and facilitates easy transportation.

Corrosion

This is the solvent action of sea water. This is of relatively minor importance since the sea water has high salt content and calcium carbonate. The chemical action of the sea water is, however, seen where coasts are composed of readily soluble rocks such as limestone, dolomite etc.

Features Formed By Marine Erosion

Erosional processes by seas produce coastal cliffs, wave-cut benches etc.

(i) *Cliffs* As we know, steep shores suffer more destruction. Cliffs are steep wall like features carved in to a bedrock mass by a combination of marine undercutting and subaerial processes. These are most important especially on coasts facing the full force of major storm waves. Because of the abrasion taking place actively at the base of the cliff, its upper part is undermined and crumbles and the cliff gradually recedes with a steep face towards the advancing sea.

Intense wave erosion causes the formation of a notch at the base of some cliffs. These are known as 'wave-cut notch' which mark the initial stages of under—cutting that leads to a mass movement as the cliff breaks away above the notch and falls into the sea.

Other minor erosional features associated with cliffs are the stacks, sea caves and sea arches. These are the isolated remnants of former land, produced as a part of the cliff retreat along lines of weakness in the rocks. These may be due to the differences in the strength of the rock, and structure of the rocks exposed.

When the rocks exposed in the cliff are composed of stratified sediments of varying strength, the weaker units are under cut rapidly forming sea caves and arches, which have resistant capping layers. When an arch collapses, columnar masses or rock remain standing isolated as islands just off the shore. These columnar rock masses are called stacks.

These are temporary features which are ultimately leveled and are of minor relevance to the overall development of a coast.

(ii) *Wavecut bench* It is known variously as abrasion platform, shore platform, marine terraces etc. With continuous abrasion the rocks that have been dislodged from the cliff get dumped at the cliff foot, which is a gently seaward sloping seafloor, called the wave-cut bench. This sloping rock floor, with continued abrasion grows wider in the landward direction.

TRANSPORTATION

The products of erosion by sea water and those which are dumped into the sea by the rivers are carried to various depths by the movement of the sea water. When the current is strong and the waves reach the bottom of the sea all the clastic material get stirred up. These are transported by rolling at low velocities and may pass in to a suspended state in high velocity.

The waves and currents play a substantial role in the transportation and distribution of the sediments whereby the concentration of coarser matrial is located closer to the shoreline and the more finegrained material are found progressively towards the central parts. Material which are soluble in sea water are transported chemically in a dissolved state.

DEPOSITION

Marine basins are the main depositional sites where most of the material obtained by the wear and tear of the rockmasses on the land through various natural agencies like river, glacier, wind etc. accumulate. Besides, the material produced through the processes of erosion by the sea also form a part of the marine deposits. Apart from the fragmental material derived from the land, the skeletal remains of the organisms dwelling in the sea basin as well as the material transported in the form of true and colloidal solutions are also involved in the formation of marine deposits. Thus marine sediments are derived from the following sources :

(a) detrital material of immediate terrigenous origin (i.e. material derived from the land).

(b) products of subaerial and submarine volcanism;

(c) inorganic precipitates from the sea water;

(d) products of chemical transformations taking place in the sea;

(e) skeletal remains of organisms and organic matter;

(f) extra-terrestrial materials (meteoric dust, meteorites etc.)

The waves and currents have a general tendency to move sediment seaward from the shore and to deposit coarser particles near the shore and finer sediments more in the seaward direction. This type of separation and sorting of the grains according to their size and specific gravity are effectively carried out by the joint action of waves and currents. While the waves move the rock-waste towards the shore (i.e. the upswash), the returning undertow current drags them towards the sea. This to and fro motion of the materials give rise to well-sorted marine sediments. In certain zones of marine deposits, a definite sequence of accumulation is noticed, which is as follows:—

Terrigenous, Biogenic, Chemogenic, Volcanic and Polygenic.

Depending on the distance of the deposits from the shore, the depth of a basin, the movement of the water, the organic activities, etc, marine deposits are of various character. Accordingly they are classified into four main groups, such as:

1. Littoral deposits.
2. Neritic or Shallow-water deposits.
3. Bathyal or Deep-sea deposits.
4. Abyssal or Very-deep water deposits.

1. *Littoral Deposits* These are the shore-zone deposits. The littoral zone includes the area between the extreme levels of the high-tides and low-tides. This zone is submerged during high tides and emerge during low tides and the deposits here are found to be unstable with respect to time. Its width may vary depending on the nature and structure of the coast. Due to the proximity of this zone to the land, from which the materials are derived, the deposits are mainly terrigenous and consist primarily of boulders, gravel, sand and mud. Mollusca, crustacea and echinoids are found mixed in these materials.

Depositional shore features include beaches, longshore bars, spits, beach barriers, barrier island and tombolos.

(a) *Beaches* As defined by Longwell and Flint (Introduction of Physical Geology, 2nd Edition, John Wiley & Sons, Inc., New York) a beach is "the wave-washed sediments along a coast extending throughout the surf zone". As we know most of the sediments entering the shore-zone of the breaking waves are mainly brought by the streams and are derived from the wave erosion or breaking down of rock units exposed along the shore. Besides, glaciers, wind and volcanoes may also contribute to these sediments only locally. The accumulation of these material between the submerged wave-cut bench and the marine cliff, gives rise to a beach.

Beaches are composed of unconsolidated materials ranging in size from pebbles to sand and mud. The sediments in a beach tend to shift along a shore as well as back and forth as the wave advances and retreats. Thus any given beach tends to change its dimension from one season to the next. This impermanence is characteristic of beach deposits.

(b) *Bars* These are ridge-like deposits formed along the coast and are composed of sand, pebble and shell accumulations. Some times they are hundreds of kilometres long, 20-30 km wide and as high as 15-30 metres. Bars often confine a portion of the sea, separting it form the main body of water. This separated part of the sea is called a lagoon. When the bar lies far offshore and is completely detached from the main land, it is known as an off shore bar. It is also termed as beach barrier or barrier island.

The term 'bar' is mostly used for those depositional features which are generally submerged below sea-level and emerge at low tides.

(c) *Spits* These are often considered as the most common types of bar. These are ridge-shaped deposits of sand and gravel which projects out from the land mass in to the sea. Thus its one end is attached to the land and the other end terminates amidst the sea. Sometimes spits may reach a considerable length.

(d) *Tombolo* It is a form of bar which connects the main land mass with an island or sometimes connects the neighbouring islands. These are also known as 'connecting bars'.

As far as the beach barriers and barrier islands are concerned, they are elongated sand ridges generally parallel to the shore, often encloses a lagoon between the ridge and the shore. Behind the beach barriers in the near-shore zone conditions are sometimes favourable for the development of peat-bogs. Besides, in some places of the littoral zone there are accumulations of fractured shells of various sea-organisms.

2. *Neritic deposits* These are the shallow water deposits formed on the continental shelf and at similar depths in the flanking regions of oceanic islands. In this zone, the sediments which are found are of terrigenous, chemogenic and biogenic nature.

By far the greater part of the deposits on the continental-shelf zone are composed of terrigenous sediments. Due to intense wave action which causes sorting of the grains according to size, the coarser fragments are laid down close to the shore, whereas the finer particles are carried away to a considerable distance out to sea. Accordingly, we find rudaceous sediments (psephites) consisting of boulders, pebbles, cobbles, gravels etc. followed by arenaceous sediments (psammites) consisting mainly of sands and then silty sediments and finally argillaceous sediments (pellites) where clayey particles predominate.

In addition to the terrigenous sediments, the neritic deposits also contain chemicalsediments. As we know, various mineral substances migrate from the land in to the marine basins solely as solutions, and precipitate under favourable physico-chemical conditions giving rise to calcareous, ferruginous, manganous and sediments of other compounds. Mineral matter brought in to the marine basins in colloidal solutions get coagulated by the electrolytes of the sea water and drop out of the solution and precipitate forming chemical sediments. Thus in the neritic zone, deposits of calcium carbonate, iron and manganese hydroxides are found.

Along with terrigenous and chemogenic sediments, the shelf-zone deposits also contain biogenic sediments. An abundance of nutrients and availability of sufficient light in the shelf-zone create favourable conditions for the development of organic life. The neritic deposits, therefore, contain organic shells and skeletons of various animals. These organisms mostly assimilate calcium carbonate (lime) dissolved in the sea water to build their skeletal parts. Deep-water neritic deposits occur around the oceanic islands. Here, the coral reefs are of special importance.

Coral Reefs

These are the best examples of rock formation by the direct action of living organisms. These are masses of limestone and dolomite, which are essentially the remains of calcareous organisms and algae. In their form, the corals resemble branching trees and shurbs.

The coral reefs are accumulated by coral polyps which live in colonies. Apart from the polyps a large number of organisms take part in building up the coral reef. Calcareous algae are as important as the corals themselves and much of the deposit consists of the shells of Foraminifera, Molluscs, Echinoderms and other creatures. Corals may not even make up the largest part of the reef. More often the term 'bioherm' is used for such structures.

Each coral polyp has a calcareous skeleton, in the form of a small tubular chamber, secreted from the animal's body. Living in colonies of millions of individuals, they combine to build a structure of great size. In reef-building corals, the successive generation of polyps settles in and fastens itself to the dead ones of the preceding generation and the reef gradually grows upward and sidewise. Thus, in a reef while the upper part is inhabited by the living polyps, the lower part is hard calcareous skeleton.

Coral polyps need, for their growth, a temperature of about 20°C. They live within depths of about 80 metres where sunlight is abundant. Fresh, muddy and highly saline waters are unfavourable to their growth. Because of temperature , light and oxygen requirements of the organisms that inhabit them, reefs are built only at or close to sea level. Coral reefs occur in warm, tropical and equatorial waters between the latitudes 30°N and 25°S. As corals can not survive long exposure to the air, living reefs cannot grow much above low-tide level and are remarkably flat on top. In recent years a few coral reefs have been found in polar seas also. Coral reefs are particularly numerous in the Pacific and Indian oceans.

On the basis of their forms and relation to the land three main types of reefs are recognised: (i) *fringing reefs,* (ii) *barrier reefs,* (iii) *atolls.*

Fringing Reefs

These are narrow belts of reef which grow around island or along the coast. The reef is closely attached to the coast and the island. At low tide they are seen to be in continuity with the shore. Its seaward edge is higher than the landward portions. It is separated from the coast by a narrow, shallow strip of water.

Barrier Reefs

These are located at a considerable distance from the coast or the island and are separated from the mainland by a quite broad and relatively deep stretch of sea. This lagoon may be tens of kilometres wide. Small channels cut across the barrier connect the lagoon with the open sea. The great Barrier Reef of Australia is the best example of this type. It extends for more than 2000 kms with an average width of 150 kms.

Atolls

These are circular coral Reefs with a shallow central lagoon but no central island. A large number of channels cutting across the atoll reef connect the lagoon with the open sea. Atolls are more common in the Pacific than in any other ocean.

Origin of Coral-Reefs

The standstill theory, the glacial control theory and the subsidence theory-have been propounded to explain the origin of coral reefs. Of these three theories, the subsidence theory by Charles Darwin is still considered valid by most geologists. According to this theory, the island around which the fringing reef develops subsides slowly, while the corals continue to build upward and outward. Thus over a period of time the original fringing reef becomes barrier reef and the island

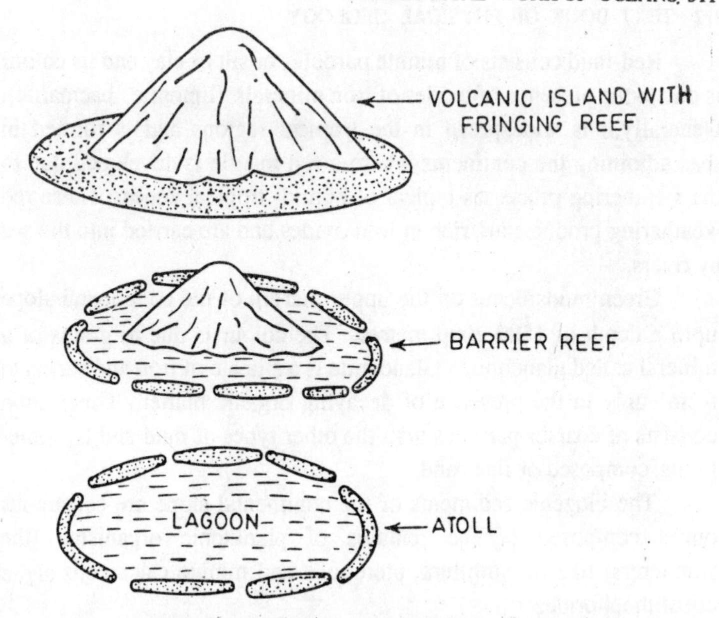

FIG.78. TYPE OF CORAL REEFS

is separated from it by a lagoon. With further subsidence the island disappears below the water level and the surrounding reef is termed as atoll.

3. *Bathyal or Deep-sea Deposits* These deposits are formed in the continental slope zone which is characterised by low mobility of the sea water. Accordingly there is least mechanical transport. Besides, the absence of light and the prevalence of low temperature in this zone makes it unfavourable for benthic animals. Most of the organisms, here are planktons with carbonate or siliceous skeleton.

The terrigenous sediments of this zone are mainly derived from the shelf-zone and consist mostly of silty-pelitic muds. According to the environment of deposition the muds may be blue, red and green. The blue colour of the mud and the smell of hydrogen sulphide indicate that its formation is in a reducing environment (i.e. deficiency of oxygen and abundance of organic matter). It is found in the zone of continental slope and adjoining parts of the ocean-floor and may extend upto depths of about 5000 metres.

Red-mud consists of minute particles of silt or clay and its colour is due to the presence of oxides of iron minerals (limonite, haematite). Generally,it is widespread in the tropical regions and is formed in seas adjoining the continents where a red mantle is developed due to the weathering processes typical of humid tropical zones. These red weathering products are rich in iron oxides and are carried into the sea by rivers.

Green muds occur on the upper margin of the continental slope upto a depth of 1500-2000 metres. The colour is due to grains of a mineral called glauconite. Glauconite is a silicate of iron and seems to form only in the presence of decaying organic matter. Green mud consists of coarser particles than the other types of mud and is, sometimes, composed of fine sand.

The biogenic sediments of the continental slope are calcareous oozes composed of the remains of planktonic organisms (the wanderers) like foraminifera, pteropoda and minute calcareous algae-cocolithophorides.

The foraminiferal oozes is composed of the minute skeletons of different foraminifera which are of calcareous nature. Pteropod ooze contain floating mulluscs called pteropoda (these are known as sea-butterflies) and accumulate generally at a depth of 3000 metres. Coccolith ooze contain a large number of coccoliths and rhabdoliths.

4. *Abyssal Deposits* Beyond the foot of the continental slope, there is very little deposits of terrigenous material, and the abyssal plain is covered for most of the part by pelagic deposits which are formed to a great extent of the shells and skeletons of animals and plants which when alive float on the surface of the water. Pelagic deposits are not entirely composed of the remain of animals and plants but also contain volcanogenic and polygenic sediments. The pelagic deposits are classified into two groups - inorganic and organic. The inorganic group is represented by the polygenic and volcanogenic sediments whereas the organic or biogenic group comprises calcareous and siliceous oozes.

Polygenic Sediments

They include red clay which is formed in those parts of the ocean which are most remote from the shore and at depths of 6000 to 8000 metres. In appearance and composition it is a true clay, consisting mainly of hydrated silicate of aluminium, coloured brown or brick-red by oxides of iron and manganese. The red clay is the predominant type of the pelagic deposits. It covers about 36 per cent of the oceanic area.

The components forming red clay are—

(i) terrigenous, fine clayey and other colloidal particles brought by the rivers and carried by currents to far off parts of the ocean;

(ii) aeolian dust;

(iii) wind-borne volcanic ash and pumice fragments floating on the waves to far off places before sinking;

(iv) iceberg-borne, materials;

(v) meteoric dust' (i.e. extra-terrestrial materials) consisting of tiny globules of nickel, iron;

(vi) Volcanic material from submarine eruptions;

(vii) Insoluble remains of foraminifers and those of *shark's* teeth and ear bones (otoliths) of whales.

On the surface of the red clays there is wide development of concretions composed mainly of manganese and iron oxides. These concretions consist of thin layers, concentrically enveloping finer meteoric particles, fragments of shells, volcanogenic particles etc. Apart from iron and manganese the elements which are of greater significance in the concretions are nickel, cobalt and copper.

Volcanogenic Sediments

Submarine volcanism ejects much of the pyroclastic material into the atmosphere by gas explosion and a large amount of these products accumulate in the vicinity of the volcanoes which leads to the formation of the volcanic oceanic islands.

The biogenic deposits which are more significant in the abyssal zone comprise calcareous and siliceous oozes. These oozes are rather different from those of the bathyal zone. These oozes, are of three principal types—Globigerina ooze, Radiolarian ooze and Diatomaceous ooze.

Globigerina Ooze

This is the most wide-spread variety of oozes as foraminifera abound in the seas of both tropical and temperate regions. Globigerina is the most abundant genus of forminifera. The globigerina ooze is made up chiefly of the calcareous shells of that organism. Globigerina ooze is most characteristically developed at depths ranging from 2700 to 4500 metres. In further deep zones calcareous sediments are not deposited, since at that depth the ocean cold waters are markedly undersaturated with calcium carbonate and the calcareous shells of the forminifers are dissolved before deposition.

Globigerina ooze is white, slightly yellow or pinkish in colour.

Radiolarian Ooze

It consists of accumulation of siliceoue shells of radiolaria, a minute organism of the phylum 'Protozoa'. It occurs at depths from 4500 to 8000 metres. It is confined to tropical seas and occurs chiefly in the tropical zones of the Pacific and Indian oceans. They are the deepest abyssal organic deposits. Rocks formed such deposits are called radiolarite.

Diatomaceous Ooze

Diatom oozes are produced by plankton plants which are microscopic algae with siliceous skeleton. Diatoms live in quite shallow water which is penetrated by sunlight. Diatomaceous oozes are mainly accumulated in the cold, near-pole regions at depths between 1000 and 6000 metres. As we approach polar regions, the globigerina and radiolarian oozes gradually decline with more and more predominance of diatomaceous oozes.

As described above, the oceans and seas carry out erosion, transportation and deposition. Sometimes the coastal regions show a remarkable degree of erosion and barriers known as 'groynes' are constructed to check the coastal erosion.

26

GEOLOGICAL WORK OF LAKES

Lakes are inland bodies of water found in natural depressions surrounded by higher ground. They vary in size from a pond to larger ones of hundred square kilometres in area and may contain either fresh or salt water. Lakes occupy about 2.7 million square kilometres or nearly 1.8% of the earth's surface.

In the case of lakes, they mostly occur above the mean sea level whereas the swamps are marshy lands where the water-table has just reached the land surface. Basins, even though are inland water bodies, have their bottom below the water-table always.

Formation of Lakes

Majority of the lakes are formed due to the action of various natural agencies and some are of tectonic origin. The modes of fomation of the lakes are as follows—

1. *By River Action*

(i) At the foot hill of a waterfall due to the impact of water or due to the whirling action of hard stones carried in river eddies hollows are formed below the general slope of the river bed. These hollows are small and are not of much depth. These hollows sometimes contain water to form lakes which generally come into existence after the river dies away.

(ii) As a result of rock-falls or landslides which commonly occurs in mountainous regions, the river valley may be completely blocked. Thus the flow of the river is checked giving rise to a lake in the upstream part of the valley.

(iii) Sometimes the sediments brought by the tributary river forms a bar across the main-river, thereby block its flow and gives rise to a lake.

(iv) If the bedrock is soluble, the solvent action of the river may produce a hollow of considerable dimensions and forms a lake.

(v) The depressed areas in dried up river beds, sometimes contain a sheet of water to the extent of producing a lake.

(vi) The surface of the flood-plain is uneven due to irregular deposition of silt and clay etc. in the form of ridges and the depressions left in between may contain water, even after the flood waters recede, to form lakes.

(vii) In the deltaic regions of big rivers, due to extensive branching of rivers and due to subsequent deposition many of the branches are transformed in to lakes.

(viii) Due to intensive meandering of the river, sometimes sections of the river are cut off from the main channel and form ox-bow, horse-shoe or cut-off lakes.

2. *By Action of Wind*

(i) As we know, the erosive action of wind is more pronounced in the absence of vegetation. Due to intense degree of deflation, hollows are excavated on the land surface, which may be filled with water during a storm of rain giving rise to a lake. As it occurs mostly in desert regions, permanent lakes are not formed.

(ii) Sometimes lakes are formed between sand dunes.

(iii) In desert region, basins surrounded by mountain ranges may become the site where the drainage lines converge to form lakes. Such lakes are broad, shallow and ephemeral in nature and are called *playa lake*.

3. *By Glacial Action* Lakes of glacial origin are mainly due to—

(i) the existence of pre-glacial topography;

(ii) due to glacial erosion;

(iii) due to damming or blocking by glacial or glacio-fluvial deposits;

Basins hollowed in solid rocks by glacial erosion and cirques are often occupied by small lakes (called tarns). The major glacial troughs usually contain large, elongated, trough lakes are referred to as *finger lakes*.

Piling up of morainic matter across their valleys cause the formation of lakes. The terminal moraine, as we already know, acts as a barrier and when the glacier retreats may hold water to form a lake. The kettle-holes left by melting of masses of stagnant ice often produce lakes. Material brought by the glaciers are dropped and form ridge-like deposits, across the valley, when the ice disappears and water collects behind the ridges to form lakes.

4. *By Marine Action* In coastal regions, the formation of spits and bars due to deposition of rock fragments and coarser sand particles by the wave action often confine a portion of marine water to form a lake called lagoon.

5. *By Action of Underground Water* The solvent action of ground water often leads to the formation of large depressions due to the roof collapse over great karst chambers. These depressions are characterised by their extensive size, flat bottom and basin-like shape with steep

sides. They are often filled with water forming what is known as *Poljee lakes*.

6. *Due to Tectonic Movements*

(i) Folding and faulting of sections of the earth's crust are mainly due to tectonic movements. Folding and differential faulting like tear faults or thrusts across the pre-existing river valleys very often block river to form a lake.

(ii) Blocks of the earth's crust downcast along the faults due to the stretching of the earth' s crust give rise to grabens which when occupied by water form lakes. Such are the lakes Nyasa, Tanganyika, Baikal etc.

(iii) Lakes may also result due to earthquake.

7. *Due to Volcanic Activity*

(i) Lakes are sometimes formed on the craters and calderas of extinct or dormant volcanoes (Eg. Lonar lake in Maharashtra).

(ii) As a result of damming of the pre-existing valleys by lava flows, lakes are formed.

(iii) Circular hollows occurring on the surface of a lava-flow, after it had congealed, may contain water to form lakes.

8. *Due to Organic Activity* Growth of coral reefs very often leads to the development of lagoons with the emergence of atolls.

9. *Due to Meteoritic Impact* Depression formed due to the impact of large mateorites may contain water to form lakes.

10. Artifical lakes can be created by constructing dams across river valleys.

GEOLOGICAL ACTION

As geological agents, lakes play an insignificant role in the gradation of the earth's surface than other natural agencies like river, wind, glacier, sub-surface water, seas etc. Excepting an inconspicuous degree of erosion brought about by lake water the geological action of lakes are mostly concerned with the deposition of sediments, since it acts as the depositional site for the detritus carried by the rivers and streams feeding the lake. The movement of the water in lakes occurs in the form of waves, that emerge with strong winds, currents and wind tides, both high and low. Like that of marine bodies of water, the erosion of lakes consists of the abrasion of shores and near-shore parts of the lake floor.

Different types of lacustrine sediments are formed depending on a variety of factors, such as—

(i) the size, shape and depth of the lake basin;

(ii) the geological structure and composition of the surrounding basin;

(iii) topography;

(iv) climate etc.

Lacustrine deposits are classified into three types—

(a) Terrigenous deposits which consist of the fragmental material brought in by rivers and streams and the detritus formed by lake abrasion.

(b) Chemogenic deposits formed through chemical precipitation of various salts and colloids dissolved in water.

(c) Biogenic deposits formed due to accumulation of various organisms on lake floor.

Important Lakes of India

(a) In Peninsular India—

1. Coastal lakes — Chilika lake, Pulicat lake, Kayal lake.
2. Lonar lake — (In Maharashtra).
3. Sambar lake — (In Rajsthan).

4. Dhands — (Small lakes of aeolian origin in Rajsthan).
5. Runn of kutch — (Marshy tract in Gujurat coast).

(b) In Extra-peninsula—

1. Lakes of Kashmir — Fresh water lake like Walur, Dal lake and saline lakes like Pangkong, Salt lake, Tsomoriri etc.
2. Lakes of kumaon — Lakes like Nainital and Bhimtal.

BIBLIOGRAPHY

1. Principles of Physical Geology *by* Arthur Holmes (Thomas Nelson and Sons Ltd 1975).
2. A Text Book of Geomorphology *by* Phillip G. Worcester. (Van Nostrand Reinhold Company New York—1969)
3. Principles of Geomorphology *by* William D. Thornbury (John Wiley & Sons Inc. New York Second Edition—1985)
4. Geomorphology *by* B.W. Sparks (Longman, London 1960)
5. Introduction to Physical Geology *by* Chester R. Longwell & Richard F. Flint. (John Wiley & Sons Inc. New York, London).
6. Earth and Its Mysteries *by* G.W. Tyrrel (G. Bell and Sons Ltd., London 1954).
7. Physical Geography *by* Philip Lake (The Macmillan Co. of India Ltd. 1974).
8. Modern Physical Geography *by* Arthur N. Strahler and Alan H. Strahler (John Wiley and sons, Inc. New York)
9. The Earth's Changing Surface *by* M.J. Bradshaw, A.J. Abbott & A.P. Gelsthorpe. (The English Language Book Society and Hodder and Stoughton) 1979.
10. The Surface of the Earth *by* A.L. Bloom (Prentice Hall, 1969)
11. Physical Geology *by* G. Gorshkov, A. Yakushova. (MIR Publishers. Moscow 1967)
12. General and Historical Geology *by* Yu. M. Vasiliev, V.S. Milnichuk and M.S. Arabaji. (MIR Publishers. Moscow 1981)
13. Geology with the elements of Geomorphology *by* A.F. Yakushova (MIR Publishers. Moscow 1986).
14. Eath Science. *by* Richard J. Ordway (Affiliated East-West Press Pvt. Ltd. New Delhi 1971).
15. The Earth Beneath Us. *by* K.F. Mather (Random House 1964)
16. Glacial and Pleistocene Geology *by* R.F. Flint (Wiley, 1957).
17. Physical Geology. *by* Sheldon Judson, Kenneth. S. Deffeyes and Robert B. Hargraves. (Prentice Hall of India Pvt. Ltd.) 1975

18. The Dynamic Earth: Text Book of Geosciences *by* Peter J. Wylli (John Wiley & Sons, Inc. New York. London. 1976)

19. Understanding the Earth. *by* I.G. Gass, Peter J. Smith, R.C. L. Wilson. (The English Language Book Society — 2nd Edition 1979).

20. Basic Concepts of Physical Geology. *by* E.W. Spencer (Oxford & IBH Publishing Co. New Delhi 1970).

21. Streams. *by* M. Morisawa (Mcgraw Hill, New York 1968).

22. Soil Conservation. *by* H.H. Bennett. (McGraw-Hill Book Co. New York and London 1947).

23. Ground Water Hydrology. *by* D.K. Todd (John Wiley and Sons, New York. 1959).

24. Marine Geology. *by* PH. H. Kuenen (John Wiley & Sons, New York, 1950).

25. Structural Geology. *by* DE. Sitter (McGraw Hill Book Co. New York. 1956).

26. Structural Geology. *by* M.P. Billings (Prentice Hall, New York 1959).

27. Elements of Structural Geology. *by* E.S. Hills. (Asia Publishing House, New Delhi).

28. Structural Geology. *by* V. Belousor. (MIR Publishers, Moscow).

29. The Sea Coast. *by* J.A. Steers. (Collins, London 1954).

30. Introduction to Geomorphology. *by* A.F. Pitty. (Methuen & Co., London. 1971).

31. The Study of Landforms. *by* R.J. Small (Cambridge University Press, Cambridge 1972)

32. The Earth's Crust and Mantle. *by* V. Meinesz. (Elsevier, New York 1964).

33. Physical Geology. *by* D.J. Leet (Prentice Hall, New York 1965).

34. Geomorphology, An introduction to the study of Landforms *by* C.A. Cotton (John Wiley and Sons, New York 1952).

35. Earth, The Living Planet. *by* M. J. Bradshaw (The English Language Book Society and Hodder and Staughton 1979).

INDEX